SINGLE SAMPLE STATISTICS
exercises in learning from just one example

ACADEMISCH PROEFSCHRIFT

ter verkrijging van de graad van doctor

aan de Universiteit van Amsterdam

op gezag van de Rector Magnificus

prof. dr. D.C. van den Boom

ten overstaan van een door het College voor Promoties ingestelde commissie,

in het openbaar te verdedigen in de Agnietenkapel

op dinsdag 31 mei 2016, te 10 uur

door

Peter Bloem

geboren te Haarlem

Promotor:	prof. dr. P.W. Adriaans	Universiteit van Amsterdam
Copromotor:	dr. S. de Rooij	Universiteit van Amsterdam
Overige leden:	prof. dr. L.F.C. Antunes	Universidade do Porto
	prof. dr. P. D. Grünwald	Centrum voor Wiskunde en Informatic
	prof. dr. F.A.H. van Harmelen	Vrije Universiteit Amsterdam
	prof. dr. ir. C.T.A.M. de Laat	Universiteit van Amsterdam
	prof. dr. A.P.J.M. Siebes	Universiteit Utrecht
	dr. J. Vreeken	Universität des Saarlandes

Faculteit der Natuurwetenschappen, Wiskunde en Informatica

The research reported in this thesis has been carried out under the auspices of SIKS, the Dutch Research School for Information and Knowledge Systems.

This research was supported by the Dutch national program COMMIT.

Contents

SUMMARY

What can be learned from a single example? If we are faced with some complex process, producing large intricate constructs, but we are only given one example of its output, can we still draw conclusions about the source, or ascribe meaning to the patterns we find? This is by no means just an academic exercise: we only have one Internet, for example, only one climate system and only one global financial system. What assumptions must we make about the processes that generated them, in order to learn about their structure? What can we do if we make no assumptions at all? Each chapter of this dissertation addresses an aspect of this question, starting with a high-level theoretical approach, and gradually working towards more practical aspects.

The first chapters provide an informal introduction to the problem, and the tools we use to study it. In **Chapter 3**, we view the problem in its most general form, using the theory of *Kolmogorov complexity*. In using Kolmogorov complexity, we make only one assumption: that the source of the data can be understood as a computational process. Under this assumption, it gives us an objective definition of the data's *information content*. As is well known, the incomputable Kolmogorov complexity can be bounded from above by computable means. We show that with additional assumptions about the source of the data, such as its computational complexity, we can compute a value that is not just an upper bound, but also, with high probability, a good approximation. We also analyze functions derived from Kolmogorov complexity, such as the *normalized information distance*: we show that good approximations to Kolmogorov complexity do not necessarily translate to good approximations of derived functions, but with careful analysis, we can provide some guarantees.

Chapter 4 deals with model selection. Given only a single sample, what can we say about the complexity of its source? How much of the data is structure, and how much is random? This question has been studied under many names, like *sophisti-*

cation, the *algorithmic sufficient statistic* and *effective complexity*. We show that all these approaches have fundamental problems: the functions proposed *cannot* correspond to the intuition that inspired them. It remains an open question whether objective model selection in this setting is possible, but we provide several arguments that suggest the answer is negative.

In **Chapter 5** we turn to a practical application of the single sample setting: large complex graphs. These are complex objects, with rich internal structure, but no straightforward way to divide the data into chunks with similar properties. One solution is to find small, frequently recurring subgraphs, known as *network motifs*. However, the fact that a subgraph is frequent is by itself no indication that it is a meaningful pattern: many subgraphs occur frequently simply by chance. To show that a particular subgraph is special, we must show that its occurrences are *unexpected* for a particular source. Using the Minimum Description Length principle, the more practical cousin of Kolmogorov complexity, we develop a fast way to judge whether such subgraphs are unexpected. This allows motif analysis to scale to much larger graphs than was possible with traditional techniques.

Where the previous chapter studies the recurrence of similar structures at the same scale, **Chapter 6** investigates *self-similarity*: the recurrence of the same structure across scales. This is often a crucial assumption in graph analysis: we cannot analyze the whole of the World Wide Web, so we assume that a large subgraph, extracted from a random walk, has the same properties as the whole. Learning self-similar structure is known as the *fractal inverse problem*, a long-standing open question. We analyze the fractal inverse problem in the domain of point patterns in Euclidean spaces, and show that it can be solved using the Expectation-Maximization algorithm.

The field of statistics is divided neatly by the type of data under analysis. The available models and techniques differ sharply from times series to sets of iid samples, to geospatial information. The single sample setting provides us with a general perspective: it shows that in all cases we are dealing with a single, finite binary string, and we are hoping to model it with a computable probability measure. The internal structure of this binary string is an assumption we make about its source, usually to let us divide the data in chunks, so that the similarities and differences between these chunks will let us reconstruct the source from the data. This view is instructive when we are faced with modern types of data like complex graphs, where the question of how to subdivide the data is not easily answered. The perspective of single sample statistics gives us a starting point: we can always consider the data a single sample from some computable distribution.

He waved a photograph of the Phaistos Disc. 'The people who made this. Four thousand years ago! They used stamps! And if they were such pre-alfabetic genuises, then surely this must say something interesting, mustn't it? So I should be the first to read it, shouldn't I? The miserable thing is we have only this one specimen. But you don't make stamps for just one tablet, do you?' [...]

'Let me explain my predicament to you.' He picked a newspaper up from the floor, and scribbled something in the margins. 'Write down the following number: eighty-three billion one hundred and ninety-one million twenty-four thousand five hundred and sixty-seven.' And after she had written it down on the sheet she used for her notes: 'Now imagine an aboriginal cryptographer in an ancient Australian forest, who doesn't even know that those are numbers; all he sees is eleven incomprehensible marks: 8 3 1 9 1 0 2 4 5 6 7. They're all different, except the repeated symbol 1. Wat can he conclude from this? Nothing at all. That's the point I'm at now. Imagine that he suddenly has the brilliant idea that they're numbers, how is he supposed to figure out that they're alphabetically ordered, Dutch numerals from one to ten? How is he supposed to figure out that "acht" is the name of the number 8? He doesn't even know the decimal system, let alone the Dutch language.

—*The Discovery of Heaven*, Harry Mulisch [69]

1 · INTRODUCTION

In the early chapters of the novel *The Discovery of Heaven* we meet Onno Quist: a disorganized, left-wing philologist, who has become obsessed with a mysterious artifact discovered centuries ago in the ruins of the Minoan palace on Crete, the *Phaistos disc*. The disc is real. It contains a spiralling sequence of 242 symbols, from an alphabet of 45 distinct signs. Quist is convinced that it holds an important message, and is sufficiently lacking in humility to decide that he must be the one to decipher it. But the strange markings that adorn the Phaistos disc are not found anywhere else. The disc is the only example of this kind of writing.

Quist is living a scientist's nightmare. Not for nothing are studies with small sample sizes easily dismissed as near-meaningless. To make strong inferences and hard claims, we cannot do without many repetitions: large amounts of examples of the same type of thing, over and over again. The recent announcement of the discovery of the Higgs Boson, inferred from experimental data with the famously stringent "five sigma" level of statistical significance that particle physics requires, took 300 trillion repetitions of the proton collision experiment that the Large Hadron Collider was built for. For a particle physicist, attempting anything with just one sample must feel like digging a railway tunnel with a toothpick.

There is still, however, considerable difference between one sample and no samples at all. Consider this classic scenario, used the world over to instruct students in the basics of statistics: a soldier for a nation engaged in a bitter ground war is debriefed after being rescued from behind enemy lines. He relates to his superiors that he witnessed a new type of tank in a training exercise. The tank is clearly a secret weapon, and far in advance of anything their side can produce. The officers become anxious and wish to know how many of these tanks the enemy posess. The soldier only saw one, but he can relate that its serial number was 17.

Assuming that the enemy number their tanks in sequence, and that the soldier was as likely to spot this tank as any other, what does this one observation tell us about the number of tanks produced? Clearly, there are at least 17, but how many beyond that? What would be a good guess? One approach would be to take the number of tanks n for which this outcome (a soldier spotting tank number 17) is most likely. For n less than 17, seeing tank 17 is impossible, and for higher n the probability of observing any particular tank is $1/n$. Thus, by this criterion we will guess that the enemy made just 17 tanks, and our soldier just happened to stumble on the one with the highest serial number. This seems like an optimistic conclusion, so perhaps we need to adjust our criterion.

Instead of choosing the n for which our particular outcome is the most likely, let us try to find a different procedure; one that (a) ensures that the average of many repeats of this experiment (one soldier, one tank) converges to the true value, and (b) produces the smallest expected difference between our guess and the true value. Such a procedure is known as an *unbiased minimum-variance estimator*. For this problem, one exists, and it tells us to expect that the enemy has 33 tanks. It also tells us that if we want to make a statement with 95% confidence, we should say that the enemy has between 17 and 340 tanks. Still a great deal of uncertainty, but much better than we would have had if we had dismissed the single sample as useless. During the second world war, the Allies used exactly this procedure to estimate the number of Mark V tanks the Germans were producing (although they had more than one observation to work with) [33].

This type of trick is not much help to Onno Quist. The Phaistos disc contains no serial number, or at least not one that we can read. And if it did, estimating the production capability of the ancient Minoans is of no more help in translating the contents of the disc, than the tank's serial number would be in guessing the names of its occupants. But Onno does have one advantage that the beleaguered nation didn't have: the Phaistos disc contains *internal structure*. We can count its symbols, look at their frequencies, their proximities and co-occurrences. The single number 17 could not be cracked open in this way to to find more information.

And in this respect Onno is not so lonely in his quest. While few scientists are asked to use just one sample to estimate the mean life expectancy, or the median income of a population, when the objects under study become more complex and richly structured, the number of examples available usually drops. Take the current webgraph, for example: the dense web of links between webpages on the Internet. This is one of the most promising "objects" of study in the world today. Unlocking just part of its structure allowed Google to completely dominate the search engine market at the end of the last century. And like Onno, researchers studying of the webgraph have only one example to work with. Climate researchers may sympathise. Until our telescopes improve their resolution, we have only one liquid-water planet whose atmosphere we can investigate. Speaking of telescopes, until 2003, astronomers had no proof that other stars even *had* planets circling them, and the only sample of a solar system they had access to was our own.

So what can be done? How far can Quist hope to get with the Phaistos disc? The world of statistics, machine learning and data mining offers a vast landscape of approaches. A landscape, that is, unfortunately, rather fragmented. The statisticians are divided in two camps, Frequentists and Bayesians, with incompatible ideas of what constitutes a probability. Furthermore, there is a zoo of different types of data: independent draws, or draws arranged in time, both in single dimensions or multiple, each draw may be from a fixed, finite number of outcomes, or from a continuous spectrum. And then there are objects like networks and trees, rich in structure, but requiring a whole new approach.

How can we capture this entire landscape in a single perspective? We will construct a metaphor to illustrate our view. Imagine a large room, filled with a vast number of machines. The machines are started, and each begins to write a sequence of ones and zeroes to a tape. Some machines will run indefinitely, and some, after a while, will stop. Each machine operates like clockwork: once started it will follow the exact same sequence of steps every time. The only thing that affects this deterministic operation is the ability the machines have to ask for randomness. The machine, at any point, as determined by its deterministic operation, can ask for a random bit. If it does, the operation of the machine pauses, we flip a coin, and provide the machine with the result: 1 if the coin lands head and 0 if it lands tails. Based on this input the machine can change its operation. Since we don't feel like standing around waiting for the machines to ask us for coin flips, we simply flip a large number of coins in advance, write the results on a large tape, and let the machine read from this tape as it pleases.

Now imagine that we are given a finite sequence of ones and zeroes, a *bit string*, and we are told only that it came from one of the machines. The question put to us, is which machine produced the data? This is a metaphor for the business of inference. The world is a collection of processes. Each partly deterministic, and partly random. Our data is some set of observations from one of these processes: perhaps a human brain formulating an order for movie tickets, perhaps a human investigator sampling the heights of randomly selected individuals. You may object that none of these datasets are bit strings, but they can all be encoded as such. In fact any statistician using a computer to analyze her data must admit that whatever shape or properties she assumes her data has, the way it is stored on the hard drive of her laptop is as a string of ones and zeroes.

Now, we haven't yet described what these machines are. How should we build them? By what rules do they operate? Can we define a family of machines so that any process that might produce our data is equivalent to some machine in our family? It turns out that there is such a family, called *Turing machines*. These are purely mathematical constructs, and although occasionally a computer scientist with time to spare will build one out of Meccano or Lego, they are mostly studied with pen and paper.

There are very good reasons to believe that any process we can hope to understand fully, is equivalent to a Turing machine. Note that we don't say that any process the universe can exhibit is necessarily Turing equivalent. We only claim that the Turing machines capture *effectiveness*. A process is *effective* when it can be described objectively and unambiguously, so that it can be simulated by anyone with access to the description. Turing machines may not capture the limits of the universe, but it is likely that they capture the limits of what we can understand (at least at the level of scientific rigor). If our data did not come from a Turing-equivalent process, we have no hope of understanding those parts that a Turing machine cannot capture. In Chapter 2, we will look at Turing machines in more detail, and discuss the reasons to believe that they capture the notion of effectiveness.

This will be our framework. We have encountered some data, and we will assume that a Turing machine (or some process equivalent to it) has produced it.[1] If we assume that we have unlimited resources at our disposal, what can we hope to do? We could test every machine: try every sequence of coin flips as input. Many machines are capable of producing our data, but for which machine are we most likely to observe our data? This is a simple matter: for every sequence of random bits for which the machine produces our data, the probability is $\frac{1}{2} \times \frac{1}{2} \times \frac{1}{2} \times \ldots$ and so on, for as many bits as we fed the machine. That is, if we fed the machine n bits, the probability of the machine producing our data in this way is 2^{-n}. If there are multiple sequences for which the machine produces our data, we can sum their probabilities. We will introduce some notation to represent these ideas concisely: let T represent one of the machines, and let y be a sequence of bits that we should feed the machine so that it produces our data x. We then write:

$$T(y) = x.$$

Note that there is always a machine that doesn't require any random bits, and simply produces our data whenever we start it. For any dataset, such a machine exists, encoding the data in the details of its cogs and wheels. For this machine, the probability of seeing our data is 1. This isn't a very satisfying explanation of our data. We would like to find a machine that will produce a dataset *like* ours when we run it a second time, but not one that produces the exact same dataset. We want a second Phaistos

---

[1] It is possible to assume that data comes from a source that cannot be simulated by a Turing machine, and still perform effective inference under this assumption. We will not discuss such settings here.

disc, not the same disc again. Somehow, we need to take into account the cost of encoding the data internally. The solution is found in *universal* machines. Somewhere in our vast room of infinite machines, there are machines that work as follows: they first choose, using the coin flips, another machine at random, in such a way that any machine can be chosen, and then proceed to *simulate* that machine.

Instead of asking which of the many machines produced our data, we can simply assume that the universal machine produced it, and ask which is the shortest sequence of random coin flips that causes the universal machine to produce our data. This sequence encodes first a machine, and then a sequence of random bits to feed to that machine to get our data. Let i be the first part of the sequence, and let y be the second. If we call the universal machine U, we can describe its operation as:

$$U(iy) = T_i(y).$$

Here, iy is the concatenation of i and y, and T_i is the machine described by i. Thus, running U and feeding it the sequence iy is equivalent to running T_i and feeding it the sequence y.

This is the basis of the theory of *Kolmogorov Complexity*. This theory is built on two important consequences of this metaphor. First, it gives us an explicit connection between how likely a dataset is, and how compactly we can describe it. Second, we can choose any universal Turing machine we like, and the Kolmogorov complexity will only change by a constant amount.

The first consequence, the connection between a *description* of our data, and its *probability*, can be shown as follows. If we take the sequence of random bits that causes our universal machine to produce our data, we can send it to somebody else, somebody who knows which universal machine we've chosen, and they can reconstruct the data from just this sequence. Thus, there is a strong connection between how compactly we can describe our data, and how likely we are to see it. This means that the most likely way for the universal machine to have produced our data is equivalent to the shortest description for the data using the universal Turing machine. And that value is called the *Kolmogorov complexity*.

The Kolmogorov complexity answers the question "how much information does an object contain?" If I can describe an object in 500 bits, then the information contained in the object cannot be more than 500 bits. So, if I find the shortest possible description, the length of that description is the amount of information contained in the object.

To make this intuition precise, we must detail what we mean by a *description*. The objects themselves, we will assume, are encoded as bit strings, so that all descriptions are of bitstrings. As for the description language, we will not prescribe what it should be, save that it is (a) effective, and (b) Turing complete. Effectiveness simply means that there is a formal, unambiguous method to get from a description to the object being described. In modern terms: there is an algorithm for unpacking the description. Turing completeness means that the description language is as powerful as the family of Turing machines. The simplest way to satisfy these requirements is to choose a universal Turing machine, and to take the sequence of bits we feed it as a description of the data it outputs. We will refer to the bits fed to the universal Turing machine as the *program* p. As before, we write the operation of U on p to produce x as:

$$U(p) = x.$$

The Kolmogorov complexity of x is the length of the shortest *program* p that we can feed U so that it produces x:

$$K(x) = \min \{|p| : U(p) = x\}$$

where $|p|$ denotes the length of p.

You may object at this point that this definition of information content is very dependent on a lot of arbitrary choices. What if we had chosen a different universal machine? Or a different description language altogether? Wouldn't we get an entirely different Kolmogorov complexity for the same x? How then, can we talk as if this *information content* is somehow a property of the data? This brings us to the second important property of Kolmogorov complexity: if we change our description language, we may indeed get a different complexity, but there's a strong bound on how much the complexity will differ. For two different description languages, the Kolmogorov complexity will differ by at most a constant amount, independent of x.

The argument is simple. First we note that any description language that fits the criteria outlined above, can be implemented by a universal Turing machine. So the question can be reduced to *how much will the Kolmogorov complexity change, if we switch to a different universal Turing machine?* The next step follows from the fact that a universal Turing machine can simulate any other Turing machine. Remember that p is the concatenation of two bit strings: one, i, describing a Turing machine, and another y, describing the bits to be fed to that Turing machine. If we denote the Turing machine described by i as T_i, then we can rewrite the operation of U as

$$U(iy) = T_i(y).$$

Now, if somebody else, with a different universal Turing machine V, claims that the Kolmogorov complexity is 500 bits, how much can we disagree with him on the basis of our universal Turing machine U? Since U can simulate any other Turing machine, it can also simulate V. Let v be a description of the machine V so that

$$U(vy) = V(y).$$

Then, one program we have to produce x, is v, concatenated with the 500-bit program that our opponent claims to have. So our Kolmogorov complexity must be less than 500 plus the length of v. Using the same argument in reverse, there is some string u such that their Kolmogorov complexity will always be less than ours plus $|u|$.

We say that Kolmogorov complexity is independent of the choice of universal Turing machine in an *asymptotic sense*: for small datasets there may be meaningful differences, but so long as the datasets grow large enough, the difference becomes insignificant. This kind of asymptotic thinking takes some getting used to, and while it is tempting to simply think of Kolmogorov complexity as an objective function per se, it is important to make such simplifications with the eyes wide open. For instance, given a dataset, we can always choose two universal Turing machines such that their "constant of disagreement" is much larger than the size of the data. However, given these two universal Turing machines, there is always a number n such that for other datasets, larger than n, the constant is dwarfed by the size of the data. To summarize, information content is subjective, but that subjectivity is bounded by a constant, while the amount of information contained in a dataset is not.

The second thing that makes Kolmogorov complexity challenging to work with is that it is incomputable. There exists no computer program (or Turing machine) that can compute it for us. The crux of the problem is that if we decide to simply try all programs for the universal Turing machine one by one, and see which produces x, there will be some programs that take a long time to finish, and some programs that never halt. We may try programs in parallel, of course, using multiple copies of U, and at any point we will have a current shortest program p, and several shorter programs that are still running. The problem is, we can never be sure if those programs still running might, at some point, stop and produce x, of if they will never halt, and we have in fact found the shortest program already.

This is the point where the practically-minded statistician may depart. An incomputable function that is only objectively defined in an asymptotic sense? What possible use is such a thing to those of us with practical, everyday jobs to do, like decoding the Phaistos disc? The point we wish to make here is not so much that we must all change our ways, to the glorious path of Kolmogorov complexity, but that Kolmogorov complexity offers a framework, a perspective on the things that we are all doing already. We all sit behind computers, so we all analyze bit strings. We all try to fit probabilistic models to these bitstrings, models that are usually equivalent to Turing machines.

And it is a perspective that shows us our limits as well as our options: the constant machine-dependence discussed above does not go away if we forget about Turing machines, even if we're just trying to fit a normal distribution to this year's exam results. We may limit our model class to normal distributions, but that just means we are restricting our view to a particular subset of Turing machines: those computing discrete approximations of normal distributions. And outside this subset of Turing machines, the incomputable ideal description still exists. Chapter 3 investigates this idea further, asking if we can find some sort of guarantee that the model class we have limited ourselves to contains a Turing machine that describes our data as well as the Kolmogorov complexity.

The incomputability of the ideal description length is one of the limits that Kolmogorov complexity imposes. Are there more? What else can't we do? If we get enough data, can we recover, with some level of certainty, the exact Turing machine that produced our data? Or will there always be multiple Turing machines that are equally likely to be responsible? We study this question in Chapter 4, and find that there are strong reasons to believe that arriving at a single choice of Turing machine is impossible. We can use the theory of Kolmogorov complexity to reject many Turing machines as the source of our data, but we will always be left with a set of equivalent models.

And what of the statisticians whose datasets do not fit well-established forms like independent draws from a single source. What can Onno Quist and researchers studying the web graph or the climate system take from Kolmogorov complexity? How can this incomputable quantity help them to cut up their data into manageable chunks and expose its inner structure? What if we find some structure, and it helps us to compress the data, can we take this as proof that the patterns we have found are somehow intrinsic to the data? In Chapter 5, we take on one such practical problem with the principles of Kolmogorov complexity: the analysis of complex graphs. We find that the notion of compression as learning helps us to find a fast method to search for such chunks in out data.

Kolmogorov complexity shows us our options, but also our limitations. As we will see, the answers to the questions posed above are negative as often as they are positive. Some of the things we do every day, with constrained model classes, become impossible if the model class grows. Others survive the generalization, and remain valid techniques, even if we extend our model class to include all Turing machines. Ultimately, this leaves us with a highly robust subset of statistical techniques. Techniques that are still applicable, no matter how powerful our model class becomes.

The arrival of global networking, exascale data storage, and petascale processing has propelled us into a new era of data analysis. An era where there is no shortage of data, but also no shortage of new *types* of data. The familiar forms of data: iid samples and timeseries, have been joined by new forms that, while large and rich in structure, show no easy way to get from the full bit string to smaller, independent components. We study two examples of such structures: in Chapter 5 we analyze complex graphs, and in Chapter 6, we study data with *self-similar* structure; for which a small part has the same properties as the whole.

Meanwhile, the model classes that are available for such novel types of data are equally diverse, and difficult to generalize. Often, the only common denominator is that they are distributions that provide a high probability for the data: or equivalently, description methods that provide a short description. This is the general perspective that Kolmogorov complexity provides: a single sample, represented by a bit string, and a universe of models, represented by Turing machines.

2 · SETTING THE STAGE

This chapter provides an informal introduction to the concepts that recur in later chapters. It is not based on original research and can be safely skipped. All chapters are self-contained, and can be understood with a basic understanding of the preliminaries. However, this chapter may help to provide context to the results presented hereafter, and allow the reader to better see the thematic throughline discussed in the introduction and conclusion.

This dissertation describes four separate research projects; each with their own context and their own aims. There are however, recurring themes and subjects. The three most important ones are *effectiveness*, *Kolmogorov complexity* and the *no-hypercompression inequality*. We will take some time to discuss each informally.

2.1 Effectiveness

Towards the end of the 19th century, mathematicians began to think about the limits of mathematics. As mathematics grew more formal, and more precise, people started to consider what the limits of such formal systems were. More and more, people began to ask about the existence of *effective procedures* to solve various mathematical problems. "Effective procedure" was a term with no mathematical definition, but one that was easily understood intuitively: a procedure that was unambiguous, requiring no insight or intuition. A set of instructions that anybody could follow to arrive at a certain outcome. What today, we would call an *algorithm*.

Here's an example: imagine you are faced with a stack of magazines, which you would like to sort by date. Move trough the stack from top to bottom. For every magazine and the one below it, check if they are in the correct order. If not, swap them around, otherwise move on. When you get to the bottom of the stack, go back to the top and repeat the procedure. If you make it to the bottom without performing a single swap, the stack is sorted. This is an *effective procedure* for sorting a stack of magazines. Note that we are not saying anything about the *efficiency* of the procedure (there are much better ways to sort a stack of magazines), just that it can be described in unambiguous instructions, that any competent individual can follow.

The quest for the existence of effective procedures is exemplified in the *Entscheidungsproblem,* posed by David Hilbert in 1928. He asked for an effective procedure that takes as its input a statement in first-order logic and decides whether or not the statement is true (that is, whether there exists a proof for the statement). The Entscheidungsproblem asks for an effective procedure for the business of being a mathematician. A way to arrive automatically at a proof, or disproof, of any properly formalized statement.

In order to get anywhere near a solution of the Entscheidingsproblem, what was needed was a formalization of the idea of an effective procedure. Some language or formalism, such that any possible effective procedure could be captured by it. The solution came when a young mathematician named Alan Turing went for a walk.

As he lay down in Grantchester meadows, near Cambridge where he was a fellow of King's College, he considered the problem of *computable numbers*. One of the first things we learn in mathematics is the distinction between different types of numbers. To start with, we have the natural numbers: 0, 1, 2, ..., the numbers we can use to count objects. Then there are the *integers*, which include the natural numbers, but also their negatives. There are the *rational numbers*: those that can be represented as a/b, where a and b are integers. There are the *algebraic numbers*: those numbers for which a polynomial function with rational numbers for coefficients is zero. Some numbers that we know, like $\pi = 3.141592\ldots$ don't belong to any of these classes. There is an infinite sequence of decimals that represents it, but no division of two integers, and no rational-valued polynomial can be used to describe it precisely. Yet, in a sense, a mathematician could "compute" π: given large stack of paper, and a sufficient amount of time, a competent mathematician could follow a simple set of instructions and write down an arbitrary number of these decimals. Is this true for all numbers, Turing wondered, or are there numbers that can *only* be represented as an infinite sequence of decimals, where even a competent mathematician, with a set of clear instructions and unlimited time is not sufficient to pin the number down.

This comes down to effective procedures. A computable number is one for which we have an effective procedure to write down the first n decimals, for any n. To answer this question, Turing needed a formal definition of what constitutes an effective description. A language for a set of instructions that is simple enough that our competent mathematician can follow them, but that is complex enough to capture anything we might consider "computation".

Turing proceeded by taking the only example we have of a system that can undeniably execute any effective procedure: a mathematician at a desk with an unlimited supply of paper, time and coffee. He then broke down this system into its essential components. Let's start with the paper. The mathematician can write whatever he likes, wherever he likes on the paper, but that's not strictly necessary. We can segment the paper into cells, and allow him to write only one symbol into each cell. This does not fundamentally restrict what he can do. It makes things more difficult, but the possibilities should remain the same. While we're at it, we can take these cells and string them into a single paper tape, along which the mathematician can only move left or right. We also require that he can only read or write to the cell right in front of him. Again, things become more laborious, but the set of things the mathematician can do, the numbers he can, in theory, compute, should remain the same. As for the symbols, we can restrict the mathematician to zeroes and ones. He can simply use small sequences of these to encode whatever other symbols he used originally, so again, while we're making his life more difficult, we are not restricting the things he can, in principle, achieve.

But of course the desk, the paper, the symbols, these are not the complicated parts. The true complexity lies in the brain of the mathematician. While we have failed, and will fail for a long time to come, to find a formal description of human intelligence, we can take a shortcut here to make our life easier. Let's assume that in the course of carrying out his instructions, our mathematician's brain can only take on a finite number of *states*. We don't know how to define the state of a brain, but we'll simply say that once a brain is in the same state, given the same context, it will always do the same thing. Recall that the mathematician is carrying out strict instructions, requiring no special insight or intelligence. Most likely, we usually won't need nearly as many states as the human brain is capable of taking on. We can even remove the mathematician's memory, save for the part that stores his instructions, since anything the mathematician needs to remember, he can write down on the tape.

By now, we have reduced the whole system to a very simple and understandable machine, known these days as a *Turing machine*. The machine moves its read/write head along an unlimited tape, reading and writing ones and zeroes on it. A *program* for such a machine consists simply of (a) how many states the machines should reserve, (b) a set of rules, with each rule telling the machine that if it is in a certain state, and reads a certain symbol, the machine should either move left, move right or write a particular thing on the tape. After that, the rule gives us a new state to move to. If we have really succeeded in reducing the mathematician step by step to a formal system, without removing any potential capability, then for any effective procedure we can find a set of rules to program such a machine with, to make it execute the procedure. The convention is to identify the machine with the program: that is, if we talk about a certain Turing machine, we are actually referring to a Turing machine with a specific program.

This gave Turing an immediate answer to the question of computable numbers: since we can sort all Turing machines in a long list (shortest first, and then alphabetically if the lengths are the same), we can *number* them 0, 1, 2, 3 and so on. This tells us that there are as many Turing machines, and hence as many computable numbers, as there are natural numbers. And since it had been known for decades that the set of natural numbers is in a specific sense much smaller than the set of all numbers that can be represented by infinite sequences of decimals, the same holds for computable numbers.

But Turing had not just solved the question of computable numbers. He had stumbled on a definition of *computability* that was to form the foundation of the field of computer science. All this, incidentally, he did well before the invention of the digital computer.

Incidentally, while our construction of the Turing machine *suggests* that we have reduced a system capable of all effective computation to a formal machine, without reducing the limit of its capabilities, this does not constitute a *proof* that every effective procedure corresponds to a Turing machine. Such a proof cannot exist, as effectiveness is an intuitive notion, but in the eight decades since the Turing machine was invented, every formal machine that has ever been imagined has been shown to be either equivalent to a Turing machine (ie. it leads to the same set of computable numbers), or weaker (it leads to a subset of the Turing-computable numbers). The hypothesis that this will remain the case, and no effective procedure exists that cannot be implemented by a Turing machine, is known as the *Church-Turing thesis*: anything that can be effectively computed, can be computed by a Turing machine.

From Turing's construction, we can also see why there must be universal Turing machines: machines that can simulate any other machine, as mentioned in the introduction. The execution of a Turing machine clearly follows a pre-determined, unambiguous process. A process that a mathematician at a desk can easily work through, given some coffee and a stack of paper. In short, there is an effective procedure for the simulation of Turing machines. And thus, if the Church-Turing thesis holds, there is a Turing machine that simulates other Turing machines. Turing did not trust to the Church-Turing thesis, however, but provided a rigorous, constructive proof that a universal Turing machine exists [88].

We can now see why, in the introduction, we chose the metaphor of a room full of machines to represent the business of statistical inference. Under the Church-Turing thesis, the set of all Turing machines captures all possible effective procedures. We have implicitly equated a *model* for a dataset with an effective procedure for generating it. To complete this picture we need to address two issues.

First, it should be noted that Turing machines as currently defined are entirely deterministic. They either produce the data or not. In the introduction we fed the random bits upon request, to introduce randomness. The simplest way to formalize this is to give the machine an input tape (separate from its work tape), from which it can only read, and whose head can only move in one direction. We fill this tape with random bits (taking care to add a few more every time the machine is in danger of reading an empty square). This provides the machine with a source of randomness and the ability to generate *samples* from a computable distribution.

Incidentally, these kinds of Turing machines, where the input tape can only be read and then only in one direction, are called *prefix-free Turing machines*. They are called this because no input that causes the machine to halt can be a prefix of another input that causes the machine to halt: if a is a prefix of b, and the machine halts if we put a on the input tape, then putting b on the input tape would have the same effect: once the machine has read the part of b that is equal to a, it will halt.

Second, our perspective uses *generative* probability models: machines that produce random samples on request. In many areas of statistics it is more common to start with models which gives numeric *probabilities*, either to outcomes, or sets of outcomes. It can be shown that the two formalisms are equivalent in the following sense: given a Turing machine that samples from a distribution, we can provide another Turing machine that approximates the probability of each outcome from below up to arbitrary accuracy. Likewise, given such a Turing machine approximating the probability, we can provide one that samples outcomes with the correct probabilities.[1]

We will conclude this section with the most famous result about Turing machines: the resolution of the Entscheidungsproblem. Turing, in the same paper in which he introduced his machines, reduced the Entscheidungsproblem to what has become known as the *halting problem*: does there exist an effective

---

[1]A proof is provided in Section A.0.1.

procedure to decide whether any given Turing machine, with a given input, either halts or computes indefinitely? Turing showed that if no such procedure exists, then an effective procedure for the Entscheidungsproblem also cannot exist. He then showed that, indeed, the halting problem could not be solved by effective means.

To see why the halting problem is *undecidable,* imagine, as we did before, the Turing machines enumerated in a long list so that each Turing machine gets its own number. To the right of each Turing machine we write what the output is, if we provide it with the input 1, encoded as a binary string, next to that the output for input 2, and so on. If the Turing machine doesn't halt for a particular input, we write ∞. This gives us a large table that extends infinitely far down, and infinitely far to the right. A table that depicts all effectively computable functions from one natural number to another.

Now consider the following procedure with input i:

Take the computation of the Turing machine numbered i, on input i. If it halts and produces a number, output that number plus 1. If it does not halt, output 0.

Is this procedure effective? If it is, then there must be a Turing machine which computes it, and this Turing machine must correspond to some row in our infinite table. Yet, by its construction, we can see that every Turing machine in our table differs from this one in at least one place: at the diagonal of our table. Thus, this procedure cannot be effective. The problem lies in the condition "if it halts". This cannot be effectively computed. The halting problem is undecidable, and with it the Entscheidungsproblem.

2.2 Kolmogorov complexity

We have met the Kolmogorov complexity in the introduction: the shortest program for x on a universal Turing machine as a formal definition of the "amount of information" that x contains. Above, we fleshed out the definition of Turing machines a little bit. We saw that the Turing machines capture all effective methods, so that we can now say the Kolmogorov complexity of x is just the length of the shortest effective method for producing x.

We also mentioned that Kolmogorov complexity is incomputable. Or, as we can now put it: there is no effective method of computing the Kolmogorov complexity. Even though it is perfectly well-defined—one program for U, after all, must be the first that produces x—it cannot be computed. How can this be?

Incomputability   Gregory Chaitin, one of Kolmogorov complexity's three independent inventors, arrived at Kolmogorov complexity in the way that is most directly related to its incomputability. He was considering the *Berry paradox*, and trying to find a resolution. To explain the Berry paradox in modern terms, imagine two mathematicians playing a game on Twitter. They are each trying to name the largest possible number within the 140 characters that Twitter allows. The first might come out with a simple:

99999999999999999999999999999999999999999999999
99999999999999999999999999999999999999999999999
99999999999999999999999999999999999999999999999

To which the other might reply with

9^9^9^9^9^9^9^9^9^9^9^9^9^9^9^9^9^9^9^9^9^9^9^9^9^
9^9^9^9^9^9^9^9^9^9^9^9^9^9^9^9^9^9^9^9^9^9^9^9^9^
9^9^9^9^9^9^9^9^9^9^9^9^9^9^9^9^9^9^9^9

a so-called *power-tower*: we raise nine to the power of nine, we raise nine to the power of that and so one. Using a more obscure description, the other, inspired by a popular webcomic,[2] comes back with:

A(G, G), with A the Ackermann function and G Graham's number.

A famous large number, entered into a fast-growing function. The second mathematician thinks for a while, and deals the final blow:

The smallest number not expressible in a tweet.

The price paid for winning the game is the birth of a paradox. The winning mathematician has just expressed a number in a tweet, that by its very definition is not expressible in a tweet. This is a modern version of the original Berry paradox, which reads *The smallest number not expressible in less than 20 words.*

---

[2]http://xkcd.com/207

Chaitin co-invented Kolmogorov complexity to solve this paradox. Translated to the world of Turing machines the description becomes a program

"Compute $K(x)$ for all x increasing in length. When you find an x for which $K(x) > ?$, stop and output x"

where some natural number n should take the place of the question mark. The size of the program grows with our choice of n, but only very slowly, so that we can be sure that for large enough n the length of the program itself is much less than n. So by the construction of the program, we have $K(x) > n$. But the program itself prints x. It is a description of x much smaller than n. By the Church-Turing thesis, we can translate this program to the universal Turing machine. If we accept that that program will also be much shorter than n, so long as n is big enough, we have arrived at our paradox. By definition, x has large Kolmogorov complexity, but we have also described it concisely.

The resolution is that our program *cannot* be implemented on a universal Turing machine. There is no effective method to execute the instruction "compute $K(x)$", and thereby, no effective method to compute our program. You might argue that we can simply make a huge number of copies of the machine T_u and run every program shorter than the length of x in parallel, one machine per program. Surely that is an effective way to compute $K(x)$? The problem with this approach is that some machines may run for a very long time and halt, while others will never halt. At any point, we will have no idea which machines, of those still running, are going to halt at some point on the future, and which machines will never halt. This problem is incomputable, and by extension, so is the value of $K(x)$.

Randomness   Finally, we will consider the solution that Kolmogorov complexity offers to another classical problem, a clash of intuition and theory that has bothered mathematicians for a long time. Imagine you are passing the the time with a friend by betting pennies on the outcome of a coin flip. Your friend flips the coin, and you take turns betting on the outcome. Heads always nets you a penny, tails costs you one. Writing '1' for heads and '0' for tails, the outcome of the first forty bets looks like this:

$$0101010101010101010101010101010101010101 \, .$$

Since you took the first bet, you have now lost twenty pennies, and you can take it no longer. You accuse your friend of cheating. Your friend counters that you have no basis for your claim: the probability of this sequence is $\frac{1}{2} \times \frac{1}{2} \times \frac{1}{2} \times \ldots = \left(\frac{1}{2}\right)^{40}$, the same as any other sequence of 40 outcomes. If it had been

$$0100101001001001001110111011010010111010 \, ,$$

would you have complained? Because both sequences have the same probability. And while you cannot counter his argument, you do decide to cut your losses and stop the game: you're still convinced that you've been cheated.

Is your friend right? Is there no basis to claim that the first bit string requires more of an explanation than a fair coin flip? The problem has existed since at least 1812, when the ideas of classical probability theory were first gathered together by Pierre-Simon Laplace. His thoughts on the issue were remarkably close to the solution we have today:

"The drawing of a white ball from from an urn which a-mong a million balls, contains only one of this color, the others being black, would appear to us likewise extraordi-nary, because we form only two classes of events relative to the two colors. But the drawing of the number 475813 from an urn that contains a million numbers seems to us an ordinary event; because comparing individually the num-bers with one another without dividing them into classes, we have no reason to believe that one of them will appear sooner than the other." [59]

So who decides what these classes are? Why does "regular strings" count as a class, but we're not allowed to put the irregular string in a class by itself? For that matter, what does "regularity" mean? If someone tells us that the second string looks regular to him, how can we convince him he's wrong?

If we assume that Turing machines, with access to randomness, were responsible for the two strings above, we can begin to see a hint of a solution: assume that both strings came from a universal Turing machine. Remember that the universal Turing machine operates by first sampling another Turing machine, and then the input for that Turing machine. For the first string, there are a few options. The first is to sample a Turing machine that does nothing more than spit out its own input. This description requires 40 bits, plus the description of the Turing machine. Another is to sample the Turing machine that repeats the the sequence 01 a number of times, determined by its input. This machine requires again a fixed number of bits to describe, and its input takes however many bits it takes to represent the number 20.

Clearly the second option provides a smaller description. To solve the problem in Laplace's terms: let a class of strings be a subset of strings that can be represented efficiently by some Turing machine. That is, the input to the Turing machine required to return the class is less than the full length of the string, plus the cost of describing the Turing machine itself. In other words, any string in such a class has a Kolmogorov complexity less than the length of the string: the string is *compressible*.

How many compressible strings are there? This is a simple computation: take the set of strings of length 40, how many of them are compressible to 10 bits? There are at most 2^{10} programs of 10 bits, so the ratio of programs compressible to 10 bits is no more than $2^{10}/2^{40} = 2^{-30}$, about one in a billion. In general terms, the rule is that the proportion of strings compressible to more than n bits than their length is 2^{-n}. This function decays very rapidly: for 10 bits it's around one in one-thousand, and for 20 bits it's one in one-million, and so on. This tells us that the proportion of strings that are compressible by more than a small margin, is very small.

This is why we were so surprised to see a string with such regularity: since flipping a random coins selects a string uniformly from the set of all strings of a given length, and the highly regular strings represent an exponentially small proportion, the probability of seeing such regularity is impossibly small.

2.3   The no-hypercompression inequality

If effectiveness and Kolmogorov complexity are the leading couple in our production, then the *no-hypercompression inequality* is its main character actor. Unassuming, unsurprising and modest, a workhorse. But no less crucial to the plot than its more glamorous colleagues. It's so modest a result that its inventor is no recorded, but the name was coined in [47, p103].

Let T be any prefix-free Turing machine and sample a string x from it. T may have many different programs for x. Call these y_1, y_2, y_3, ... For each y that causes T to produce x, the probability of providing T with that y is $2^{-|y|}$. Summing the probabilities over all possible programs for x, the total probability that T produces x, is

$$p_T(x) = 2^{-|y_1|} + 2^{-|y_2|} + 2^{-|y_3|} + \dots$$

Let $L_T(x) = -\lceil \log p_T(x) \rceil$. In effect, L_T combines the probability mass of all the programs y_1, y_2, y_3, ... into a single codelength.

The no-hypercompression inequality tells us that if we sample from T, then with overwhelming probability K(x) is not much less than $L_T(x)$. Specifically, the probability that we sample a string such that $L_T(x) - K(x)$ is more than k bits is 2^{-k} bits. Again, we see an exponential decay, meaning that for only 30 bits difference, the probability is already below one in a billion.

As an example, consider the Turing machine, that simply copies the first 400 random bits it samples to its output. This machine is equivalent to simply flipping a random coin to generate a bit string. The probability that you will flip a string compressible by 30 bits is less than one in a billion. This is exactly what we already saw in the last section.

Since the Kolmogorov complexity is more efficient, up to a constant, than any effective description method, it follows that if we sample x from T the probability that *any* other Turing machine can be used to describe x more efficiently than L_T can, by a more than k bits, decays exponentially in k. Indeed, it turns out

---

[2] See Section A.0.1.

we need not worry about the "up to a constant" this time. It can be shown that the no-hypercompression inequality holds for the negative logarithm of any probability distribution: for any probability distribution p and any description method D, if we sample x from p, the probability that D describes x more efficiently than L_T by k bits is less than 2^{-k}. [3]

The no-hypercompression inequality has many uses. First, it tells us that if we assume that our data came from T, and we approximate the Kolmogorov complexity with L_T, which is computable for most reasonable T, we can be almost certain that we have, up to a few bits, an accurate approximation of the Kolmogorov complexity, even though the Kolmogorov complexity itself is not computable. We expand on this idea in Chapter 3, showing that even if we broaden our assumption to state that one in a *set* of Turing machines produced the data, we can still apply this principle, and arrive at a solid, computable approximation of the Kolmogorov complexity.

In Chapter 5 we use another logical consequence of the no-hypercompression inequality: if we assume that T generated x, and we find a compression method that compresses x better than T by k bits, we must either accept that we have witnessed a very rare event (of probability less than 2^{-k}), or reject our assumption that T generated the data. This is known as a *hypothesis test*, a common tool in statistical analysis.

As an example, consider a researcher, like Onno with the Phaistos disc, faced with a single bit string and no means to make any assumptions about its origin. And say that the researcher has a hunch that maybe it would be a good idea to consider successive chunks of five bits as the "words" of the data. Without making assumptions about the source of the data, she can't confirm that this is the case, but she can reject other hypotheses. For instance, she can make the assumption that each bit is independently drawn, from some distribution giving 0 and 1 each a probability: a kind of coin flip with an unbalanced coin. This assumption corresponds to a Turing machine, and thus a code-length for the data. If, by cutting the data into chunks of *five*

---

[3]This more generic form is shown in Chapter 5. Its proof is very simple, but requires some basic information theory that would overburden this chapter.

bits, she can compress better than this codelength, that gives her evidence to reject the assumption that the data consists of independently drawn bits.

source: http://xkcd.com/505

3 · A SAFE APPROXIMATION OF KOLMOGOROV COMPLEXITY

The material in this chapter was adapted from the paper A safe approximation of Kolmogorov complexity. ***P. Bloem**, F. Mota, S. de Rooij, L. Antunes and P. Adriaans* Algorithmic Learning Theory 2014, 336-350

As we discussed in the last two chapters, any compression we can find for a given dataset is an upperbound for the Kolmogorov complexity. Let's say Onno takes the Phaistos disc, suitably digitized, and runs it through a popular compressor, like ZIP. This fits into our perspective in the following way: somewhere in our enumeration of Turing machines, there is one, let's call it T_{zip}, that implements the unzipping algorithm: it reads a zipped file from its input tape, and spits out the unzipped version. This means that the pair T_{zip} and y form a program on the universal Turing machine for the Phaistos disc. There may be other programs, so the Kolmogorov complexity might be smaller than the cost of describing T_{zip} and x, but not longer: any effective description forms an upper bound.

Still, there is no certainty that our upper bound is at all near the mark. Consider the following sequence of digits:

1196387334243375281763971529445208602609820423210781885383640346955237554824114081209821556429859089428076534624542387362109946869363814426813302041177480603581

This will look entirely random to all but a handful of people. Put it through zip, or any other modern day compressor, and you will not be able to represent it any more any efficiently than just writing it down. Nevertheless, it's highly compressible: these are the first 160 odd places in the the decimal sequence of π. π is a well-studied number, and very efficient algorithms exist for enumerating its digits. One such program, combined with instructions to disregard the even places and to stop after a 160 characters suffices as an explanation. In short, we were fooled: we thought the string contained no structure, when in fact there is a very short description.

In this chapter, we show that at the cost of one assumption, we can show that such situations are unlikely to occur. The assumption takes the form of a subset of all Turing machines, a *model class* C. We assume that the source of our data was equivalent to a Turing machine in C. For many choices of C, we have a computable approximation of the Kolmogorov complexity, that is very rarely very wrong. Specifically, if we sample data x from any Turing machine in C, and compress it with $\overline{\kappa}^C$, our computable C-based approximation, the probability that $\overline{\kappa}^C(x) - K(x)$ is larger than k bits, decays exponentially.

We can see k as a margin of error. If the probability that $\overline{\kappa}^C(x) < K(C)$ is bounded by some c, and we need to set our margin of error at k bits to ensure a probability of less than $\frac{1}{2}c$ that our approximation is within k bits of the real value, then, with 2k bits we get a probability below $\frac{1}{4}c$. With 10k bits we get less than $\frac{1}{1024}c$ and with 20k bits we get $\frac{1}{1048576}c$, a probability of less than one-millionth of c. This result follows almost directly from the no-hypercompression inequality dicussed in the last chapter, but to a achieve a truly computable safe approximation, some care must be taken, and we we see that some obvious choices turn out not to be safe approximations.

Of course, our margin of error is usually not up for discussion, but it *is* usually proportional to the amount of data. Getting the complexity wrong by a handful of bits when the original data was only fifty bits long may be a problem, but when we have several gigabytes of data to analyze, a handful will hardly be noticeable. So, instead of imagining the statistician increasing her margin of error until the probability is low enough, we can imagine her increasing the amount of data. A twenty-fold increase in the amount of data is no mean feat, but in the era of "big data" it is certainly achievable. And if the payoff is an almost certainly accurate approximation of the magical, incomputable Kolmogorov complexity, it may be a price worth paying. We call such functions *safe approximations*.

3.1   The two worlds of Kolmogorov complexity

The Kolmogorov complexity of an object is its shortest description, considering all computable descriptions. It has been described as "the accepted absolute measure of information content of an individual object" [36], and its investigation has spawned a slew of derived functions and analytical tools. Most of these tend to separate neatly into one of two categories: the platonic and the practical.

On the platonic side, we find such tools as the normalized information distance [61], algorithmic statistics [36] and sophistication [92, 8]. These subjects all deal with incomputable "ideal" functions: they optimize over all computable functions, but they cannot be computed themselves.

To construct practical applications (i.e. runnable computer programs), the most common approach is to take one of these platonic, incomputable functions, derived from Kolmogorov complexity (K), and to approximate it by swapping K out for a computable compressor like GZIP [38]. This approach has proved effective in the case of normalized information distance (NID) [61] and its approximation, the normalized compression distance (NCD) [27]. Unfortunately, the switch to a general-purpose compressor leaves an analytical gap. We know that the compressor serves as an upper bound to K—up to a constant—but we do not know the difference between the two, and how this error affects the error of derived functions like the NCD. This can cause serious contradictions. For instance, the normalized information distance has been shown to be non-approximable [85], yet the NCD has proved its merit empirically [27]. Why this should be the case, and when this approach may fail has, to our knowledge, not yet been investigated.

We aim to provide the first tools to bridge this gap. We will define a computable function which can be said to approximate Kolmogorov complexity, with some practical limit to the error. To this end, we introduce two concepts:

- We generalize resource-bounded Kolmogorov complexity (K^t) to *model-bounded Kolmogorov complexity*, which minimizes an object's description length over any given enumerable subset of Turing machines (a *model class*). We explicitly assume that the source of the data is contained in the model class.

- We introduce a probabilistic notion of approximation. A function approximates another *safely*, under a given distribution, if the probability of them differing by more than k bits, decays at least exponentially in k. [1]

While the resource-bounded Kolmogorov complexity is computable in a technical sense, it is never computed practically. The generalization to model bounded Kolmogorov complexity creates a connection to the *Minimum Description Length* (MDL) principle [79, 80, 47], which does produce algorithms and methods that are used in a practical manner. Kolmogorov complexity has long been seen as a kind of platonic ideal which MDL approximates. Our results show that MDL is not just an upper bound to K, it also approximates it in a probabilistic sense.

Interestingly, the model-bounded Kolmogorov complexity itself—the smallest description using a single element from the model class—is not a safe approximation. We can, however, construct a computable, safe approximation by taking into account all descriptions the model class provides for the data.

The main result of this chapter is a computable function $\overline{\kappa}$ which, under a model assumption, safely approximates K (Theorem 3.3). We also investigate whether a $\overline{\kappa}$-based approximation of NID is safe, for different properties of the model class from which the data originated (Theorems 3.5, 3.6 and 3.7).

3.2 Turing machines and algorithmic probability

We will first review briefly, in technical terms, the matter that was covered informally in the previous chapters: Turing machines, computable probability distributions and Kolmogorov complexity.

Turing machines   Let $\mathbb{B} = \{0, 1\}^*$. We assume that any dataset is encoded as a finite binary string. Specifically, the natural num-

---

[1]This consideration is subject to all the normal drawbacks of asymptotic approaches. For this reason, we have foregone the use of big-O notation as much as possible, in order to make the constants and their meaning explicit.

bers can be associated to binary strings, for instance by the bijection: $(0, \epsilon)$, $(1, 0)$, $(2, 1)$, $(3, 00)$, $(4, 01)$, etc, where ϵ is the empty string. To simplify notation, we will sometimes conflate natural numbers and binary strings, implicitly using this ordering.

We fix a canonical prefix-free coding, denoted by $\bar{x}$, such that $|\bar{x}| \leqslant |x| + 2\log|x|$. See [62, Example 1.11.13] for an example. Among other things, this gives us a canonical pairing function to encode two strings x and y into one: $\bar{x}y$.

For Turing machines, we use the basic model from [62, Example 3.1.1]. The following properties are important: the machine has a read-only, right-moving input tape, an auxiliary tape which is read-only and two-way, two read-write two-way worktapes and a read-write two-way output tape.[2] All tapes are one-way infinite. If a tape head moves off the tape or reads beyond the length of the input, the machine enters an infinite loop.

For the function computed by Turing machine i on input p with auxiliary input y, we write $T_i(p \mid y)$ and $T_i(p) = T_i(p \mid \epsilon)$. The most important consequence of this construction is that the programs for which a machine with a given auxiliary input y halts, form a prefix-free set [62, Example 3.1.1]. This allows us to interpret the machine as a probability distribution (as described in the next subsection).

We fix an effective ordering $\{T_i\}$. We call the set of all Turing machines $\mathcal{C}$. There exists a universal Turing machine, which we will call U, that has the property that $U(\bar{i}p \mid y) = T_i(p \mid y)$ [62, Theorem 3.1.1].

Probability We want to formalize the idea of a probability distribution that is *computable*: it can be simulated or computed by a computational process. For this purpose, we will interpret a given Turing machine T_q as a probability distribution p_q: each time the machine reads from the input tape, we provide it with a random bit. The Turing machine will either halt, read a finite number of bits without halting, or read an unbounded number of bits. $p_q(x)$ is the probability that this process halts and produces x: $p_q(x) = \sum_{p:T_q(p)=x} 2^{-|p|}$. We say that T_q *samples* p_q.

---

[2]Multiple work tapes are only required for proofs involving resource bounds.

Note that if p is a semimeasure, $1 - \sum_x p(x)$ corresponds to the probability that this sampling process will not halt.

We model the probability of x conditional on y by a Turing machine with y on its auxiliary tape: $p_q(x \mid y) = \sum_{p:T_q(p|y)=x} 2^{-|p|}$.

The *lower semicomputable semimeasures* [62, Chapter 4] are an alternative formalization. We show that it is equivalent to ours:

Lemma 3.1. [†] The set of probability distributions sampled by Turing machines in $\mathcal{C}$ is equivalent to the set of lower semicomputable semimeasures.

The distribution corresponding to the universal Turing machine U is called m: $m(x) = \sum_{p:U(p)=x} 2^{-|p|}$. This is known as a universal distribution. K and m dominate each other, i.e. $\exists c \forall x :$ $|K(x) - \log m(x)| < c$ [62, Theorem 4.3.3].

3.3 Model-bounded Kolmogorov complexity

In this section we present a generalization of the notion of resource-bounded Kolmogorov complexity. We first review the unbounded version:

Definition 3.1. Let $k(x \mid y) = \arg\min_{p:U(p|y)=x} |p|$. The prefix-free, conditional *Kolmogorov complexity* is

$$K(x \mid y) = |k(x \mid y)|$$

with $K(x) = K(x \mid \epsilon)$.

Due to the halting problem, K is not computable. By limiting the set of Turing machines under consideration, we can create a computable approximation.

Definition 3.2. An *effective model class* $C \subseteq \mathcal{C}$ is a computably enumerable set of Turing machines. Its members are called *models*. A *universal model* for C is a Turing machine U^C such that $U^C(\bar{i}p \mid y) = T_i(p \mid y)$ where i is an index over the elements of C.

[†]Proof in the appendix.

All model classes referred to in this chapter are effective, and we will omit the adjective for the remainder. In the next chapter we will use a more general definition of model class.

Definition 3.3. For a given C and U^C we have

$$K^C(x) = \min\{|p| : U^C(p) = x\},$$

called the *model-bounded Kolmogorov complexity*.

K^C, unlike K, depends heavily on the choice of enumeration of C. A notation like K_{U^C} or $K^{i,C}$ would express this dependence better, but for the sake of clarity we will use K^C.

We define a model-bounded variant of m as

$$m^C(x) = \sum_{p:U^C(p)=x} 2^{-|p|}$$

which dominates all distributions in C:

Lemma 3.2. For any $T_q \in C$, $m^C(x) \geqslant c_q p_q(x)$ for some c_q independent of x.

Proof.
$$m^C(x) = \sum_{i,p:U^C(\bar{i}p)=x} 2^{-|\bar{i}p|}$$

$$\geqslant \sum_{p:U^C(\bar{q}p)=x} 2^{-|\bar{q}|} 2^{-|p|} = 2^{-|\bar{q}|} p_q(x).$$ □

Unlike K and $-\log m$, K^C and $-\log m^C$ do not dominate one another. We can only show that $-\log m^C$ bounds K^C from below ($\sum_{U^C(p)=x} 2^{-|p|} > 2^{-|K^C(x)|}$). In fact, as shown in Theorem 3.1, $-\log m^C$ and K^C can differ by arbitrary amounts.

Example 3.1 (resource-bounded Kolmogorov complexity [62, Chapter 7]). Let $t(n)$ be some time-constructible function.[3] Let T_i^t be the modification of $T_i \in \mathcal{C}$ such that at any point in the computation, it halts immediately if more than k cells have been written to on the output tape and the number of steps that have passed is less than $t(k)$. In this case, whatever is on the output tape is taken as the output of the computation. If this situation does not occur, T_i runs as normal. Let $U^t(\bar{i}p) = T_i^t(p)$. We call this model class C^t. We abbreviate K^{C^t} as K^t.

Since there is no known means of simulating U^t within $t(n)$, we do not know whether $U^t \in C^t$. It can be run in $ct(n) \log t(n)$ [62, 50], so we do know that $U^t \in C^{ct \log t}$.

Other model classes include Deterministic Finite Automata, Markov Chains, or the Normal distribution (suitably discretized). These have all been thoroughly investigated in coding contexts in the field of Minimum Description Length [47].

3.4   Safe approximation

When a code-length function like K turns out to be incomputable, we may try to find a lower and upper bound, or to find a function which dominates it. Unfortunately, neither of these will help us. Such functions invariably turn out to be incomputable themselves [62, Section 2.3].

To bridge the gap between incomputable and computable functions, we require a softer notion of approximation; one which states that errors of any size may occur, but that the larger errors are so unlikely, that they can be safely ignored:

Definition 3.4. Let f and f_a be two functions. We take f_a to be an approximation of f. We call the approximation b-*safe (from above)* for a distribution (or *adversary*) p if for all k and some $c > 0$:

$$p(f_a(x) - f(x) \geq k) \leq cb^{-k}.$$

Since we focus on code-length functions, usually omit "from above". A *safe* function is b-safe for some $b > 1$. An approximation is safe for a model class C if it is safe for all p_q with $T_q \in C$.

While the definition requires this property to hold for all k, it actually suffices to show that it holds for k above a constant, as we can freely scale c:

Lemma 3.3. If $\exists_c \forall_{k:k>k_0} : p(f_a(x) - f(x) \geq k) \leq cb^{-k}$, then f_a is b-safe for f against p.

---

[3]I.e. $t : \mathbb{N} \to \mathbb{N}$ and t can be computed in $O(t(n))$ [15].

Proof. First, we name the k below k_0 for which the ratio between the bound and the probability is the greatest:

$$k_m = \arg\max_{k \in [0,k_0]} \left[p(f_a(x) - f(x) \geqslant k)/cb^{-k} \right] .$$

We also define $b_m = cb^{-k_m}$ and $p_m = p(f_a(x) - f(x) \geqslant k_m)$. At k_m, we have $p(f_a(x) - f(x) \geqslant k_m) = p_m = \frac{p_m}{b_m} cb^{-k_m}$. In other words, the bound $c'b^{-k}$ with $c' = \frac{p_m}{b_m} c$ bounds p at k_m, the point where it diverges the most from the old bound. Therefore, it must bound it at all other $k > 0$ as well. □

This asymptotic definition requires some justification. After all, if the exponential decay only holds after some constant, how can we be sure that the safety of an approximation will actually come into play for our data? A similar problem occurs when analysing the time-complexity of algorithms: we may prove that a sorting algorithm has a log-linear time complexity in an asymptotic sense, but can we be sure that this will hold for some given dataset, or does the log-linear regime only start for much larger data?

The reason we resort to an asymptotic definition of safety is the same as it is with sorting algorithms: we want our claim to be machine-independent. We want any claims of safety to be independent of the indexing chosen for the model class C. Changing this index will cause a constant change to the approximation we will define later. Similarly, changing the indexing of $\mathcal{C}$ (as captured by our choice of universal Turing machine) will change the value of the Kolmogorov complexity by a constant. Since we want to make statements about safe approximation independent of such choices, we must resort to an asymptotic definition.

We choose exponential decay as the measure of safe approximation because this is the strongest decay for which we can prove that a computable function exists. We do not expect that computable functions can be found for a stronger level of decay, but this is an open question.

Safe approximation, domination and lowerbounding form a hierarchy:

Lemma 3.4. Let f_a and f be code-length functions. If f_a is a lower bound on f, it also dominates f. If f_a dominates f, it is also a safe approximation.

Proof. Domination means that for all x: $f_a(x) - f(x) < c$, if f_a is a lower bound, $c = 0$. If f_a dominates f we have $\forall p, k > c :$ $p(f_a(x) - f(x) \geqslant k) = 0$. □

Finally, we show that safe approximation is transitive, so we can chain together proofs of safe approximation; if we have several functions with each safe for the next, we know that the first is also safe for the last.

Lemma 3.5. The property of safety is transitive over the space of functions from $\mathbb{B}$ to $\mathbb{B}$ for a fixed adversary.

Proof. Let f, g and h be functions such that

$$p(f(x) - g(x) \geqslant k) \leqslant c_1 b_1^{-k} \text{ and}$$

$$p(g(x) - h(x) \geqslant k) \leqslant c_2 b_2^{-k}.$$

We need to show that $p(f(x) - h(x) \geqslant k)$ decays exponentially with k. We start with

$$p\left(f(x) - g(x) \geqslant k \vee g(x) - h(x) \geqslant k\right) \leqslant c_1 b_1^{-k} + c_2 b_2^{-k}.$$

$\{x : f(x) - h(x) \geqslant 2k\} \subseteq \{x : f(x) - g(x) \geqslant k \vee g(x) - h(x) \geqslant k\}$, so that the probability of the first set is less than that of the second set: $p\left(f(x) - h(x) \geqslant 2k\right) \leqslant c_1 b_1^{-k} + c_2 b_2^{-k}$. Which gives us

$$p\left(f(x) - h(x) \geqslant 2k\right) \leqslant c b^{-k}$$

$$p\left(f(x) - h(x) \geqslant k'\right) \leqslant c b'^{-k'}$$

with $b = \min(b_1, b_2)$, $c = \max(c_1, c_2)$ and $b' = \sqrt{b}$. □

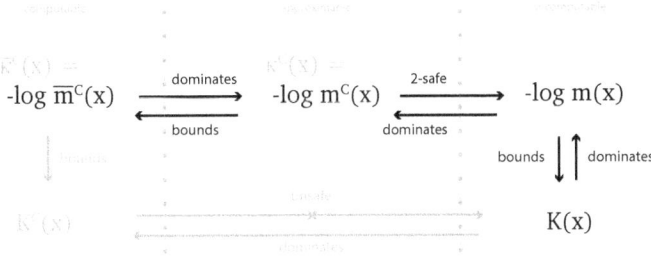

Figure 3.1: An overview of how various codelength functions relate to each other in terms of approximation safety. These relations hold under the assumption that the data is generated by a distribution in C and that C is sufficient and complete.

3.5 A safe, computable approximation of K

Assuming that our data is produced from a model in C, can we construct a computable function which is safe for K? An obvious first choice is K^C. For it to be computable, we would normally ensure that all programs for all models in C halt. Since the halting programs form a prefix-free set, this is impossible. There is however a property for prefix-free functions that is analogous. We call this *sufficiency*:

Definition 3.5. A sufficient model T is a model for which every infinite binary string contains a halting program as a prefix. A *sufficient model class* contains only sufficient models.

We can therefore enumerate all inputs for U^C from short to long in series to find $k^C(x)$, so long as C is sufficient. For each input, U^C either halts or attempts to read beyond the length of the input.

In certain cases, we also require that C can represent all $x \in \mathbb{B}$ (i.e. $m^C(x)$ is never 0). We call this property *completeness*:

Definition 3.6. A model class C is called *complete* if for any x, there is at least one p such that $U^C(p) = x$.

We can now say, for instance, that K^C is computable for sufficient C. Unfortunately, K^C turns out to be unsafe:

Theorem 3.1. There exist model classes C so that $K^C(x)$ is an unsafe approximation for $K(x)$ against some p_q with $T_q \in C$.

Proof. We first show that K^C is unsafe for $-\log m^C$.

Let C contain a single Turing machine T_q which outputs x for any input of the form $\bar{x}p$ with $|p| = x$ and computes indefinitely for all other inputs.

T_q samples from $p_q(x) = 2^{-|\bar{x}|}$, but it distributes each x's probability mass uniformly over many programs much longer than $|\bar{x}|$.

This gives us $K^C(x) = |\bar{x}| + |p| = |\bar{x}| + x$ and $-\log m^C(x) = |\bar{x}|$, so that $K^C(x) + \log m^C(x) = x$. We get

$$m^C(K^C(x) + \log m^C(x) \geqslant k) = m^C(x \geqslant k) =$$

$$\sum_{x:x \geqslant k} 2^{-|\bar{x}|} \geqslant \sum_{x:x \geqslant k} 2^{-2\log x} \geqslant k^{-2}$$

so that K^C is unsafe for $-\log m^C$.

It remains to show that this implies that K^C is unsafe for K. In Theorem 3.2, we prove that $-\log m^C$ is safe for K. Assuming that K^C is safe for K (which dominates $-\log m^C$) implies K^C is safe for $-\log m^C$, which gives us a contradiction.   □

Note that the use of a model class with a single model is for convenience only. The main requirement for K^C to be unsafe is that the prefix tree of U^C's programs distributes the probability mass for x over many programs of similar length. The greater the difference between K^C and $-\log m^C$, the greater the likelihood that K^C is unsafe. Note also that we can blow this difference up arbitrarily: for any computable function $f(x)$, we can define a model class C so that $K^C(x) + \log m^C(x) \geqslant f(x)$.

Our next candidate for a safe approximation of K is $-\log m^C$. This time, we fare better. Here, we require for the first time the *no-hypercompression inequality* [47, p103], discussed already in the previous chapter. In our current notation it reads:

Lemma 3.6. Let p_q be a probability distribution. The corresponding code-length function, $-\log p_q$, is a 2-safe approximation for any other code-length function against p_q. For any p_r and $k > 0$: $p_q(-\log p_q(x) + \log p_r(x) \geqslant k) \leqslant 2^{-k}$.

Theorem 3.2. $-\log m^C(x)$ is a 2-safe approximation of $K(x)$ against any adversary from C.

Proof. Let p_q be some adversary in C. We have

$$p_q(-\log m^C(x) - K(x) \geqslant k)$$

$$\leqslant cm^C(-\log m^C(x) - K(x) \geqslant k) \qquad \text{by Lemma 3.2,}$$

$$\leqslant c2^{-k} \qquad\qquad\qquad\qquad \text{by Lemma 3.6.} \qquad \square$$

While we have shown m^C to be safe for K, it is defined as an infinite sum, even if C is sufficient, so computing it naively would require infinite time. We can, however, define an approximation, which, for sufficient C, is computable and dominates m^C.

Definition 3.7 (Safe approximation algorithm). Let the model class D be the union of C and some arbitrary sufficient and complete distribution from $\mathcal{C}$.

Let $\overline{m}_c^C(x)$ be the function computed by the following algorithm: Dovetail the computation of all programs on $U^D(x)$ in cycles, so that in cycle n, the first n programs are simulated for one further step. After each such step we consider the probability mass s of all programs that have stopped (where each program p contributes $2^{-|p|}$), and the probability mass s_x of all programs that have stopped and produced x. We halt the dovetailing and output s_x if $s_x > 0$ and the following stop condition is met:

$$\frac{1-s}{s_x} \leqslant 2^c - 1.$$

Note that if C is sufficient, so is D, so that s goes to 1 and s_x never decreases. Since all programs halt, the stop condition must be reached. The addition of a complete model is required to ensure that s_x does not remain 0 indefinitely.

Lemma 3.7. If C is sufficient, $\overline{m}_c^C(x)$ dominates m^C with a constant multiplicative factor 2^{-c} (i.e. their code-lengths differ by at most c bits).

Proof. We will first show that $\overline{m}_c^C$ dominates m^D. Note that when the computation of $\overline{m}^C_c$ halts, we have $\overline{m}^C_c(x) = s_x$ and $m^D(x) \leqslant s_x + (1 - s)$. This gives us:

$$\frac{m^D(x)}{\overline{m}^C_c(x)} \leqslant 1 + \frac{1 - s}{s_x} \leqslant 2^c .$$

Since $C \subseteq D$, m^D dominates m^C (see Lemma A.2 in the appendix) and thus, $\overline{m}^C_c$ dominates m^C. $\qquad\square$

An alternative phrasing of this result is that $m^C(x)$ is a *computable real function* [63, Definition 4.1.2]. The parameter c in $\overline{m}^C_c$ allows us to tune the algorithm to trade off running time for a smaller constant of domination. We will usually omit it when it is not relevant to the context.

Putting all this together, we have achieved our aim:

Theorem 3.3. For a sufficient model class C, $-\log \overline{m}^C$ is a safe, computable approximation of $K(x)$ against any adversary from C.

Proof. We have shown that, under these conditions, $-\log m^C$ safely approximates $-\log m$ which dominates K, and also that $-\log \overline{m}^C$ dominates $-\log m^C$. Since domination implies safe approximation (Lemma 3.4), and safe approximation is transitive (Lemma 3.5), we have proved the theorem. $\qquad\square$

Figure 3.1 summarizes this chain of reasoning and other relations between the various code-length functions mentioned.

The negative logarithm of m^C will be our go-to approximation of K, so we will abbreviate it with κ:

Definition 3.8. $\kappa^C(x) = -\log m^C(x)$ and $\overline{\kappa}^C(x) = -\log \overline{m}^C(x)$.

Finally, if we violate our model assumption we may lose the property of safety. For adversaries outside C, we cannot be sure that κ^C is safe:

Theorem 3.4. There exist adversaries p_q with $T_q \notin C$ for which neither κ^C nor $\overline{\kappa}^C$ is a safe approximation of K.

Proof. Consider the following algorithm for sampling from a computable distribution (which we will call p_q):

- Sample $n \in \mathbb{N}$ from some distribution $s(n)$ which decays polynomially.
- Loop over all x of length n return the first x such that $\kappa^C(x) \geqslant n$.

Note that at least one such x must exist by a counting argument: if all x of length n have $-\log \overline{m}^C(x) < n$ we have a code that assigns 2^n different strings to $2^n - 1$ different codes.

For each x sampled from q, we know that $\overline{\kappa}(x) \geqslant |x|$ and $K(x) \leqslant -\log p_q(x) + c_q$. Thus:

$$p_q(\overline{\kappa}^C(x) - K(x) \geqslant k) \; \geqslant \; p_q(|x| + \log p_q(x) - c_q \geqslant k)$$

$$= p_q(|x| + \log s(|x|) - c_q \geqslant k) \; = \; \sum_{n:n+\log s(n)-c_q \geqslant k} s(n).$$

Let n_0 be the smallest n for which $2n > n + \log s(n) - c_q$. For all $k > 2n_0$ we have

$$\sum_{n:n+\log s(n)-c_q \geqslant k} s(n) \geqslant \sum_{n:2n \geqslant k} s(n) \geqslant s\left(\tfrac{1}{2}k\right). \qquad \square$$

For C^t (as in Example 3.1), we can sample the p_q constructed in the proof in $O(2^n \cdot t(n))$. Thus, we know that κ^t is *safe* for K against adversaries from C^t, and we know that it is *unsafe* against C^{2^t}.

3.6 Approximating normalized information distance

Definition 3.9 ([61, 27]). The normalized information distance between two strings x and y is

$$\mathrm{NID}(x, y) = \frac{\max\left[K(x \mid y), K(y \mid x)\right]}{\max\left[K(x), K(y)\right]}.$$

The information distance (ID) is the numerator of this function. The NID is neither lower nor upper semicomputable [85]. Here, we investigate whether we can safely approximate either function using κ. We define ID^C and NID^C as the ID and NID functions with K replaced by $\overline{\kappa}^C$. We first show that, even if the adversary only combines functions and distributions in C, ID^C may be an unsafe approximation.

Definition 3.10. A function f is a (b-safe) *model-bounded one-way function*[4] for C if it is injective, and for some b > 1, some c > 0, all q ∈ C and all k:

$$p_q \left(\kappa^C(x) - \kappa^C(x \mid f(x)) \geq k \right) \leq cb^{-k}.$$

Theorem 3.5. [†] Under the following assumptions:

- C contains a model T_0, with $p_0(x) = 2^{-|x|}s(|x|)$, with s a distribution on $\mathbb{N}$ which decays polynomially or slower,
- there exists a model-bounded one-way function f for C,
- C is *normal*, i.e. for some c and all x: $\kappa^C(x) < |\bar{x}| + c$

ID^C is an unsafe approximation for ID against an adversary T_q which samples x from p_0 and returns $\bar{x}f(x)$.

If x and y are sampled from C independently, we can prove safety:

Theorem 3.6. [†] Let T_q be a Turing machine which samples x from p_a, y from p_b and returns $\bar{x}y$. If $T_a, T_b \in C$, $ID^C(x, y)$ is a safe approximation for ID(x, y) against any such T_q.

The proof relies on two facts:

- $\bar{\kappa}^C(x \mid y)$ is safe for K(x \mid y) if x and y are generated this way.
- Maximization is a *safety preserving operation*: if we have two functions f and g with safe approximations f_a and g_a, $\max(f_a(x), g_a(x))$ safely approximates $\max(f(x), g(x))$.

For *normalized* information distance, which is dimensionless, the error k in bits as we have used it so far does not mean much. Instead, we use f/f_a as a measure of approximation error, and we introduce an additional parameter ϵ:

Theorem 3.7. [†] We can approximate NID with NID^C with the following bound:

$$p_q \left(\frac{NID(x, y)}{NID^C(x, y)} \notin \left(1 - \frac{k}{c}, 1 + \frac{k}{c} \right) \right) \leq c'b^{-k} + 2\epsilon$$

---

[4]This is similar to the Kolmogorov one-way function [14, Definition 11].

with

$$p_q(\text{ID}^C(x, y) \geqslant c) \leqslant \epsilon \text{ and } p_q\left(\max\left[\kappa^C(x), \kappa^C(y)\right] \geqslant c\right) \leqslant \epsilon$$

for some $b > 1$ and $c' > 0$, assuming that p_q samples x and y independently from models in C.

3.7 Discussion

This chapter investigated the question of what can be accomplished with minimal assumptions. With no assumptions but effectiveness, we can do no better than bound the Kolmogorov complexity from above. With a general model assumption C, however, we can show that the function $\overline{\kappa}^C$ is a safe approximation, which is computable so long as C is sufficient. We have also shown that $K^C(x)$ is not safe. Finally, we have given some insight into the conditions on C and the adversary, which can affect the safety of NCD as an approximation to NID.

Since, as shown in Example 3.1, resource-bounded Kolmogorov complexity is a variant of model-bounded Kolmogorov complexity, our results apply to K^t as well: K^t is not necessarily a safe approximation of K, even if the data can be sampled in t and κ^t *is* safe if the data can be sampled in t. Whether K^t is safe ultimately depends on whether a single shortest program dominates among the sum of all programs, as it does in the unbounded case.

For complex model classes, $\overline{\kappa}^C$ may still be impractical to compute. In such cases, we may be able to continue the chain of safe approximation proofs. For instance, we may show that a model which is only locally optimal, found by an iterative method like gradient descent, is still a safe approximation of the global optimum. By the transitive property, this would show that it is also a safe approximation of K. Such proofs would truly close the circuit between the ideal world of Kolmogorov complexity and modern statistical practice.

4 · THE PROBLEM OF SOPHISTICATION

The material in this chapter was adapted from the paper Two problems for sophistication *P. Bloem, S. de rooij, P. Adriaans* Algorithmic Learning Theory 2015, 379-394

Sophistication was proposed to complement Kolmogorov Complexity in areas where the latter does not chime with our intuition. The previous chapters have hopefully convinced you that Kolmogorov complexity formalizes very neatly the notion of how much information an object contains. Yet the objects that are the richest in information according to Kolmogorov complexity don't seem to us very interesting. They are the static you see on an old fashioned TV, the noise you hear on your car's radio when you try to find a station or simply the sequence of ones and zeroes generated by a coin-flipping game.

These are not signals that are rich or valuable to us. What makes data interesting is a mixture of predictability and unexpectedness: melodies, language, plot twists. Images of rolling hills or cityscapes. Most of these would probably have a Kolmogorov complexity somewhere in between a small constant and the full length of the raw data. However, not all data with moderate Kolmogorov complexity is interesting: flipping a coin that is slightly bent, so that it lands heads more often than tails will create an utterly boring sequence, with medium compressibility.

Is there some other way to separate the wheat from the chaff? Can we find some function, in the spirit of Kolmogorov complexity, that tells us which data is *interesting*? Sophistication is one approach: it tells us to consider not the descriptive complexity of the whole data, but of the model alone. This make intuitive sense; all the boring data mentioned above may have high complexity, but the models are all simple. Contrast this with the optimal compression of a movie: such a model should contain information about language, human anatomy, plot structure, human emotion, architecture, nature. All these concepts are patterns present in the data, and the shortest program on the universal

Turing machine exploits all of them. It must have a highly complex model.

Where the theory breaks down, as we discuss in this chapter, is the idea that this shortest program separates neatly into structure and noise: the structure placed into the Turing machine T_i and the noise placed in its input y. As we saw earlier, there are programs with the universal Turing machine for a model, that place all the information in the input, and the length of this representation is equal to the Kolmogorov complexity.

4.1 Sophistication

Kolmogorov complexity gives us a sound definition of the amount of information contained in a binary string. It does not, however, capture what most people would consider complexity. For example, a sequence of a million coin flips will almost certainly have maximal Kolmogorov complexity, even though there is nothing complex about flipping a coin repeatedly. Many scholars have defined additional measures in the spirit of Kolmogorov complexity, aimed at quantifying not *all* information in a binary string, but only the *meaningful*. While this concept has been given many names, we use *sophistication* as an umbrella term. In this chapter, we investigate two serious problems with sophistication. We conclude with two arguments suggesting the problems are fundamental, explaining our belief that sophistication cannot be defined in a satisfactory manner.

The Kolmogorov complexity $C(x)$[1] of a binary string x is, informally, the length of the shortest computer program to print x. This length depends on the choice of programming language, but, by the invariance theorem [63, Section 2.1], only by a constant, independent of x. For sufficiently complex objects, the choice of programming language becomes irrelevant and Kolmogorov complexity becomes an *objective* measure. A definition of sophistication $S(x)$ in the spirit of $C(x)$ should have similar guarantees:

---

[1] In previous chapters, we defined Kolmogorov complexity using Turing machines that can only read their input left to right, *prefix-free Turing machines*. Such Turing machines lead to the prefix-free Kolmogorov complexity $K(x)$. In this chapter, we also use the classical Kolmogorov complexity $C(x)$, defined on Turing machines that can read back an forth on their input tape at will. Since the

1. $S(x)$ should count the bits required for an effective description of the structural properties of a binary string.

2. An analogue of invariance should hold: there must be strict limits on how much sophistication can be affected by a change in programming language.

3. There should be no constant c such that $S(x) \leqslant c$ for every input x. If sophistication is bounded, then knowing its value under one programming language provides no constraints on its value under another language (except that it is also bounded).

4. Similarly, there should be no constant c such that
$$|C(x) - S(x)| \leqslant c \text{ for all } x,$$
because then sophistication would be equivalent to Kolmogorov complexity.

There have been many proposals for such a measure, all based on a *two-part code*: we encode a *model* in the first part of the code, which is interpreted as a representation of x's structural properties. The model does not fully specify x, but when combined with the second part of the code, which specifies the noise, the original string becomes fully determined.[2]

For any string x, there may be many different two-part codes. The total length can never be less than the Kolmogorov complexity, but it can come close. Figure 4.1 illustrates the principle. The key to sophistication is to take the representations that come close to the Kolmogorov complexity, the *candidates*, and define the sophistication as the size of the smallest model in this set. However, for most definitions, we can prove that they fail one of the conditions above. For others, we cannot *prove* they conflict with our requirements, but we show these methods only assign substantial sophistication to strings that require an enormous amount of processing to construct.

A valid definition of $S(x)$ must contend with two important issues. First, the details of the way the model is encoded are

_____

sophistication knows many definitions, some using $C(x)$, some using $K(x)$, we must consider both variants.

[2]Some variants deviate from the two-part coding format, see Section 4.4.3.

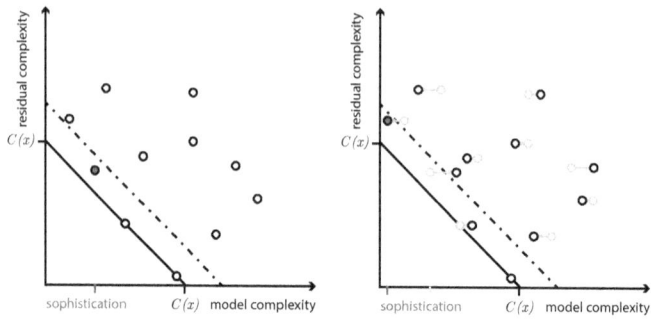

Figure 4.1: (left) Two-part representations of x by the two components of their code. The Kolmogorov complexity $C(x)$, appearing as a black diagonal, provides a lower bound on the total code-length. We consider only representations that are close to this optimum with the threshold represented by a dashed line. The size of the smallest model below the threshold is the sophistication. (right) The same image, after a constant perturbation in the model complexity caused by a change in numbering.

important. There are two technically distinct approaches; in one of these one has to deal with the so-called "nickname problem" that strangely remains unresolved in several publications. These definitions yield a sophistication that is highly dependent on the chosen programming language, unless special care is taken, as discussed in Section 4.3.

The second issue is that of striking the right balance between under- and overfitting, which we consider in Section 4.4. Overfitting is a common problem in statistics, that refers to the tendency to choose a complex model that provides a very good fit to the observed data, but does not generalise well to unseen data. In the case of sophistication, overfitting occurs if the model that determines the sophistication contains much or even all of the noise. In statistics, overfitting is often addressed by penalising complex models. In sophistication, however, such penalties tend to break the balance between structural information and noise, and lead to the opposite problem: underfitting.

Underfitting occurs when the selected model is simple, but

fails to capture all structure in the data. This is also a problem for sophistication because the models under consideration are so powerful. In particular, in any programming language, there are programs that implement an interpreter for *another* language. Such *universal models* are *simple*, since they can be described with a relatively small number of bits, yet are able to represent any data using a code within a constant from the Kolmogorov complexity. Such a two-part representation essentially encodes all information as noise. If complex models are penalized, then the problem becomes to make sure that universal models are not *always* preferred for complex data. The usual workaround is to restrict the set of allowed models, for instance to total functions. While this excludes universal models, it is questionable whether it adequately solves the problem of underfitting in general.

Finally, in Section 4.5 we argue that while two-part coding can yield useful insights into the structure of the data and identifies some models as poor representations, it is probably not possible to objectively separate structure from noise and identify a *single* model as "best": many models of different complexities may be reasonable representations. Rather than doggedly trying to "fix" this property of algorithmic statistics, we propose embracing the idea that the data allows for multiple, equivalent interpretations of which information is structured, and which is random, and that there is no such thing as sophistication.

4.2   Notation

The following notation allows us to generalize across all definitions and variants of sophistication, save the occasional exception which we will highlight individually.

In the previous chapter we dealt with Turing machines and conflated them with the functions they compute. In this chapter, we will focus more strongly on the functions themselves. A *partial computable function* is one that is computable by Turing machine. We consider non-prefix-free Turing machines as well (the prefix-free Turing machines are a subset of these).

We deal with partial computable functions $f : \mathbb{B} \times \mathbb{B} \to \mathbb{B}$, which we also call *models*. f is called *prefix* if

$$\mathrm{dom}_z(f) = \{y : f(y, z) \neq \infty\}$$

is a prefix free set for all z, i.e. no string in $\text{dom}_z(f)$ is a prefix of another. A function f is *total* if $\forall_z \text{dom}_z(f) = \mathbb{B}$. In most cases, we do not use the second argument, and let $f(x) = f(x, \epsilon)$.

A *numbering* is an enumeration of the partial computable functions, denoted by $\psi_1, \psi_2, \ldots$ or simply ψ. We fix one canonical numbering ϕ, chosen to be *effective*: i.e. given i and y, we can effectively compute $\phi_i(y)$. We call a numbering ψ *acceptable* if there exist total, computable functions $a, b : \mathbb{N} \rightarrow \mathbb{N}$ with $\forall : i$, $\phi_i = \psi_{b(i)}$ and $\psi_i = \phi_{a(i)}$. One example of an effective, acceptable numbering is to take descriptions of Turing machines (by some standard scheme), first by length, and then lexicographically. The rank in this ordering corresponds to the number in the numbering.

We generalize the definition of a model class. A *generic model class* is a set of indices in a numbering ψ. We define four classes:

- The indices of the partial computable functions $\mathcal{C} = \mathbb{N}$.
- The total functions $\mathcal{T} = \{i : \psi_i \text{ is total}\}$. Note that $\mathcal{T}$ is not computably enumerable.
- $\mathcal{K}$ is an enumerable set such that $\{\psi_i : i \in \mathcal{K}\}$ is the set of all partial computable prefix functions.
- The finite sets: $\mathcal{F}$ is an enumerable set such that $\{\psi_i : i \in \mathcal{F}\}$ is the set of uniform codes for all finite sets.[3]

The model class $\mathcal{T}$ highlights the contrast with the *effective* model classes of the previous chapter: $\mathcal{T}$ is not effectively enumerable. All model classes in this chapter are generic, and we will omit the adjective in the remainder.

For technical reasons, we deviate slightly from the traditional notation of Kolmogorov complexity used in the previous chapter: let $\mathcal{M}$ be a model class and ψ an acceptable numbering, then let $C^{\mathcal{M}, \psi}(x \mid z) = \min\{|\bar{\iota}y| : \psi_i(y, z) = x, i \in \mathcal{M}\}$, with $C^{\mathcal{M}, \psi}(x) = C^{\mathcal{M}, \psi}(x \mid \epsilon)$. We omit the numbering when the distinction is not relevant.

$C^{\mathcal{C}}(x)$ corresponds to the plain Kolmogorov complexity $C(x)$. $C^{\mathcal{K}}(x)$ corresponds to the prefix-free version $K(x)$ from the previous chapter. This is a different construction of $K(x)$, but it can

---

[3] A uniform code for set F is a surjective prefix function $f : \{0, 1\}^{\lceil \log |F| \rceil} \rightarrow F$.

be shown that the two are equal up to a constant.

Note that the notation $C^{\{i\},\psi}(x)$ can be used to represent the smallest two-part description of x using model ψ_i.

In these constructions we have used the principle of a numbering for the purpose normally served by the universal Turing machine: it captures the ad-hoc and subjective choices made in the construction of the Kolmogorov complexity, and by extension, the sophistication.

We prefer to work with numberings as it highlights an important issue: while Kolmogorov complexity is invariant to the choice of numbering this property does not immediately carry over to sophistication: for some treatments, the result is highly dependent on the chosen numbering, as we will see in the next section.

4.3 Inefficient indices

The simplest approach to sophistication would be to 'open up' the Kolmogorov complexity and to see which program achieves the smallest description length: the program that *witnesses* the Kolmogorov complexity. This witness is a two-part coding; it consists of a model and an input.

Definition 4.1 (Index sophistication). Let ψ be an acceptable numbering. Let $\mathcal{M}$ be the model class from which candidates are chosen, and let $\mathcal{N}$ be the model class that determines the minimum achievable complexity. Let c be a fixed constant. The *index sophistication* is:

$$S_{\text{index}}^{\mathcal{M},\mathcal{N},\psi,c}(x) = \min\left\{|i| : C^{\{i\},\psi}(x) \leqslant C^{\mathcal{N},\psi}(x) + c,\ i \in \mathcal{M}\right\}.$$

When $\mathcal{M} = \mathcal{N}$, we will use $S_{\text{index}}^{\mathcal{M},\psi,c}$. If the set over which the minimum is taken is empty, the sophistication is undefined.

Koppel and Atlan's treatment [57, 58], where the name *sophistication* originates, follows this basic logic, although it contains idiosyncrasies like the use of monotonic models, and an extension to infinite strings. As the subsequent history of sophistication has discarded these, we will not discuss them here.

In [13, 12] Koppel's principle is limited to finite strings, with $\mathcal{T}$ as a model class. The definition is similar to $S_{\text{index}}^{\mathcal{T},\mathcal{C},\psi,c}$, except the total complexity of a witness (i, y) is measured as $|i| + |y|$ without the cost of delimiting the two. This difference is not relevant to the current discussion. The restriction to $\mathcal{J}$ is a common approach, which avoids underfitting, as discussed in the next section.

Lemma 4.1. Let S^{ψ}_{index} denote any index sophistication with respect to numbering ψ (with any choice for $\mathcal{M}$, $\mathcal{N}$ and c). There are acceptable numberings ψ and ξ such that for all x: $|S^{\psi}_{\text{index}}(x) - S^{\xi}_{\text{index}}(x)| \geqslant \frac{1}{2} \min\{S^{\psi}_{\text{index}}(x), S^{\xi}_{\text{index}}(x)\}$.

Proof. Let $z_i \in \mathbb{B}$ consist of $2^i - 1$ zeroes followed by a one. Define ψ, ξ such that $\psi_j(x) = \phi_i(x)$ for $j = z_{2i}$ and $\xi_j(x) = \phi_i(x)$ for $j = z_{2i+1}$, with all other functions returning ∞ for all inputs. Choose any x and assume w.l.o.g. that $S^{\psi}_{\text{index}}(x) \leqslant S^{\xi}_{\text{index}}(x)$. By construction, we have $2S^{\psi}_{\text{index}}(x) \leqslant S^{\xi}_{\text{index}}(x)$. $\square$

Thus, the length of the index is a very poor indicator of model complexity. For a robust measure, we define the complexity of a function f as in [46, 91] by

$$C^{\mathcal{M}, \psi}(f) = \min\{C^{\mathcal{M}, \psi}(i) : \psi_i = f\}. \tag{4.1}$$

Lemma A.6 in the appendix shows that $C^{\mathcal{C}}(f)$ and $C^{\mathcal{K}}(f)$ are invariant.

Note that the perversely inefficient numberings of Lemma 4.1 are no issue for Kolmogorov complexity: we can use a UTM with a more efficient numbering as a model at only a constant penalty. For sophistication, however, the numbering is crucial.

There are two ways to use $C^{\mathcal{M}}(f)$ for more robust attempts to define sophistication. Confusingly, both are used in the literature. First, we can measure the complexity of the *model* ϕ_i as $C^{\mathcal{K}}(\phi_i)$, which is then the size of the first part of a two-part code describing the data. This approach is used in [31, 37, 91, 40].

Second, we can stick to using the length of the index as the measure of sophistication, but restrict the allowed numberings to those that can represent a given function *efficiently*. This approach is taken by Adriaans in [8], who defines *facticity* as $S^{\mathcal{C}, \psi, 0}_{\text{index}}$, but only allows *faithful* numberings. Formally, a faithful numbering has the property that $\forall i \exists j : \psi_i = \psi_j, |j| \leqslant C^{\mathcal{C}}(\psi_j) + c$, for some constant c. Essentially, this means that a faithful numbering can represent a function f with an index the same length as the Kolmogorov complexity $C^{\mathcal{C}}(f)$.

Contrary to Adriaans' suggestion, there do actually exist faithful, acceptable numberings:

Lemma 4.2. There are faithful acceptable numberings.

Proof. Let $d \in \mathbb{N}$ be an index such that $\phi_d(y) = \infty$ for all y. Define

$$\psi_q = \begin{cases} \phi_{\phi_i(p)} & \text{if q can be written as } \bar{i}p \text{ and } \phi_i(p) < \infty, \\ \phi_d & \text{otherwise.} \end{cases}$$

It may seem that the second line requires a test whether $\phi_i(p)$ halts, for ψ to be acceptable, but as we will show below, this is not the case.

To show that ψ is faithful, pick any function f. Then

$$C^{\mathcal{C},\phi}(f) = \min\{C^{\mathcal{C},\phi}(i) : \phi_i = f\}$$

$$= \min\{\min\{|\bar{a}b| : \phi_a(b) = i\} : \phi_i = f\}$$

$$= \min\{|\bar{a}b| : \phi_{\phi_a(b)} = f\}$$

$$= \min\{|\bar{a}b| : \psi_{\bar{a}b} = f\}.$$

This shows there is a sufficiently small ψ index.

To show that ψ is acceptable, let $\phi_j(z) = z$. Then a ϕ-index i can be mapped to a ψ-index with $r(i) = \bar{j}i$, so that $\psi_{r(i)}(y) = \psi_{\bar{j}i}(y) = \phi_i(y)$. For the reverse, define $\phi_v(\bar{i}p, y) = \phi_{\phi_i(p)}(y)$. For fixed $\bar{i}p$, the s_m^n-theorem [55] states that we can compute the h such that $\phi_h(y) = \phi_v(\bar{i}p, y)$. Let $h(\bar{i}p)$ denote this index as a function of the program; further define $h(q) = d$ if q cannot be expressed as $\bar{i}p$. By construction h is total and computable. To check that the mapping returns the correct function, rewrite $\phi_{h(\bar{i}p)}(y) = \phi_v(\bar{i}p, y) = \phi_{\phi_i(p)}(y) = \psi_{\bar{i}p}(y)$. Note that if q can be written as $\bar{i}p$, but $\phi_i(p)$ diverges, h(q) will still return a function, but one which doesn't halt, making it equivalent to ϕ_d as required. □

However, even choosing a faithful numbering is not enough. The Kolmogorov complexity uses representations of the form $\bar{\imath}y$, with $\psi_i(y) = x$, where the bar denotes some straightforward prefix encoding to delimit the model description i from its input y. If we define a second prefix encoding $\tilde{\imath}$, with $|\tilde{\imath}| - |\bar{\imath}|$ unbounded, we can define a second representation $\bar{u}\tilde{\imath}y$, with $\psi_u(\tilde{\imath}y) = \psi_i(y)$, at a constant overhead $|\bar{u}|$, and gain more than $|\bar{u}|$ for sufficiently complex strings, resulting again in a bounded sophistication.

We continue with a sophistication that avoids the issues of inefficient indices and of inefficient prefix encodings. We change the definition of index sophistication so that its two-part representations use $C^{\mathcal{K}}(\phi_i)$ bits for the representation of the model. We first introduce the following notation for the $\mathcal{M}$-Kolmogorov complexity using such compact two-part representations:

$$C_{\text{comp}}^{\mathcal{M},\psi}(x) = \min\{C^{\mathcal{K},\psi}(\psi_i) + |y| : \psi_i(y) = x, i \in \mathcal{M}\}.$$

For model classes $\mathcal{K}$ and $\mathcal{C}$ this is equivalent to the existing definition and invariant to the numbering. Note that again, we use $C_{\text{comp}}^{\{i\},\psi}(x)$ to represent the smallest two-part code using model ψ_i.

Definition 4.2 (Sophistication).

$$S^{\mathcal{M},\mathcal{N},\psi,c}(x) = \min\left\{C^{\mathcal{K}}(\phi_i) : C_{\text{comp}}^{\{i\},\psi} \leqslant C_{\text{comp}}^{\mathcal{N},\psi}(x) + c, i \in \mathcal{M}\right\}.$$

4.4 Balancing under- and overfitting

In the last section, we began to see the delicate balance between the two code components. We will study this balance, starting with the variant $S^{\mathcal{K},\psi,c}$, which is not used in the literature, but helps to illustrate the issues we wish to discuss.

$\mathcal{K}$ has optimal representations with all but a constant part of the information in the input and it has optimal representations with all information in the model. The downside to this balance is that it becomes easy to show a lack of invariance. We can tweak the numbering so that models in a specific subset $\mathcal{M}' \subset \mathcal{K}$ become cheaper to represent by an arbitrary amount relative to others: we can ensure that a model in $\mathcal{M}'$ always determines the sophistication. For instance, if we let $\mathcal{M}'$ contain only a universal model we get a bounded sophistication.

Theorem 4.1 (Underfitting). Let $\mathcal{M}, \mathcal{N}$ be model classes with $\mathcal{M} \subseteq \mathcal{N}$ and let $\mathcal{M}$ contain a universal model ϕ_u, with the property that $\exists c \forall i \in \mathcal{N}, x \in \mathbb{B} : C_{comp}^{\{u\},\phi}(x) \leqslant C_{comp}^{\mathcal{N},\phi}(x) + c$. Then, for some numbering ψ, $S^{\mathcal{M},\mathcal{N},\psi,c}$ is bounded.

This problem is well known and many treatments avoid it by restricting the model class. Less well known, perhaps, is that the same holds in the other direction: if $\mathcal{M}'$ is the set of *singleton* models—those models that output a single x for an empty input—we get a sophistication equal to the Kolmogorov complexity.

Theorem 4.2 (Overfitting). Let $\mathcal{X} \subseteq \mathbb{B}$. Let $\mathcal{M} \subseteq \mathcal{N} \subseteq \mathcal{K}$ be model classes where for every $x \in \mathcal{X}$ there is a singleton model $i \in \mathcal{M}$ with $\phi_i(\epsilon) = x$. Then there is a numbering ψ, and a constant c, such that for all $x \in \mathcal{X}$ we have $C^{\mathcal{K}}(x) - S^{\mathcal{M},\mathcal{N},\psi,c}(x) \leqslant c$.

The proofs of both theorems rely on a simple principle: there exist numberings which have the effect of penalizing $C^{\mathcal{K}}(\phi_i)$ for any model outside $\mathcal{M}'$ by an arbitrary constant amount. We can use this to effectively 'push' these models outside of the range of candidates, ensuring that, under this numbering, a model in $\mathcal{M}'$ always determines the sophistication. The requirements for $\mathcal{M}'$ are somewhat complex. The following lemma gives a set of sufficient conditions.

Lemma 4.3. Let $\mathcal{M}$ and $\mathcal{N}$ be any model class, let $\mathcal{X}$ be any set of binary strings and let $D : \mathbb{B} \to \mathbb{N}$ be a partial computable decoding function with a prefix-free domain that maps function descriptions to their indices in ϕ. Let $\mathcal{M}' = \text{range}(D)$. Further assume there is a constant c such that:

(a) $\forall_{m \in \mathcal{M}'} : \min\{|p| : \phi_{D(p)} = \phi_m\} \leqslant C^{\mathcal{K},\phi}(\phi_m) + c$

(b) $\forall_{x \in \mathcal{X}} : C_{comp}^{\mathcal{M}',\phi}(x) - C_{comp}^{\mathcal{N},\phi}(x) \leqslant c$.

Then, there is a ψ such that if $S^{\mathcal{M},\mathcal{N},\psi,k}(x)$ is defined, then $S^{\mathcal{M},\mathcal{N},\psi,k}(x) = S^{\mathcal{M}',\mathcal{N},\psi,k}(x)$ up to a constant.

Proof. Pick any $x \in \mathcal{X}$. Let f and g be ϕ-indices such that $f \in \mathcal{M}'$ and $g \notin \mathcal{M}'$ nor is ϕ_g equivalent to any function indexed by $\mathcal{M}'$. Furthermore let $C_{comp}^{\{f\},\phi}(x)$ and $C_{comp}^{\{g\},\phi}(x)$ both be within a constant q of $C_{comp}^{\mathcal{N},\phi}(x)$. Assumption (b) ensures that $\mathcal{M}'$ always provides such an f.

We will show that for every integer r, there is a numbering ψ such that $C_{comp}^{\{g'\},\psi}(x) - C_{comp}^{\{f'\},\psi}(x) \geqslant r$ for all $x \in X$, where f' and g' are the ψ-indices equivalent to f and g. Thus, for large enough r, ϕ_g is eliminated as a candidate model, while ϕ_f remains in place. Thus, under ψ, a member of $\mathcal{M}'$ determines the sophistication, or the sophistication is undefined.

Let d be a positive constant. We will show later how to choose it to achieve the required result. We define ψ as follows:

$$\psi_0(p) = 0^d 1 D(p)$$

$$\psi_{0^d 1 i}(p) = \phi_i(p)$$

$$\psi_j(\cdot) = \infty \text{ if } j \neq 0 \text{ and } j \neq 0^d 1 \ldots$$

The key to the proof is the way that the function complexity $C^{\mathcal{K}}(\cdot)$ changes when we change the numbering from ϕ to ψ. For f, the value increases by no more than a fixed constant, but for g, it increases by a constant that we can arbitrarily increase by increasing d.

We will first show that for f, the value does not increase by more than a constant c_f. Assume w.l.o.g. that $0 \in \mathcal{K}$.

$$C^{\mathcal{K},\psi}(\phi_f) = \min\left\{ |\bar{j}q| : \psi_{\psi_j(q)} = \phi_f, j \in \mathcal{K} \right\} \qquad \text{rewriting (4.1)}$$

$$\leqslant \min\left\{ |\bar{0}q| : \psi_{\psi_0(q)} = \phi_f \right\} \qquad \text{choose } j = 0$$

$$= \min\left\{ |q| : \phi_{D(q)} = \phi_f \right\} + |\bar{0}|$$

$$\leqslant C^{\mathcal{K},\phi}(\phi_f) + c_f \qquad \text{by assumption (a).}$$

In order to show that for g, we can increase the difference by an arbitrary constant, we first show that, for any z not in the range of ψ_0, the Kolmogorov complexity itself increases by at least d when we switch from ϕ to ψ:

$$C^{\mathcal{K},\psi}(z) = \min\{|\bar{i}y| : \psi_i(y) = z, i \in \mathcal{K}\} \qquad \text{by definition}$$

$$= \min\left\{|\overline{0^d 1j}y| : \psi_{0^d 1j}(y) = z\right\} \quad \text{since } z \notin \text{range}(\psi_0)$$

$$\geqslant \min\{|\bar{j}y| : \phi_j(y) = z\} + d$$

$$= C^{\mathcal{K},\phi}(z) + d. \tag{4.2}$$

We now show the increase in model complexity for g. First, assume $\phi_g \neq \psi_0$:

$$C^{\mathcal{K},\psi}(\phi_g) = \min\left\{C^{\mathcal{K},\psi}(i) : \psi_i = \phi_g\right\}$$

$$= \min\left\{C^{\mathcal{K},\psi}(0^d 1j) : \phi_j = \phi_g\right\}$$

$$= C^{\mathcal{K},\psi}(0^d 1j)$$

$$\geqslant C^{\mathcal{K},\phi}(0^d 1j) + d \quad \text{by (4.2)}$$

$$\geqslant C^{\mathcal{K},\phi}(j) - c_g + d \quad \text{since } C^{\mathcal{K}}(j) \leqslant C^{\mathcal{K}}(0^d 1j) + c_0$$

$$\geqslant C^{\mathcal{K},\phi}(\phi_g) - c_g + d.$$

Now assume $\phi_g = \psi_0$. We have

$$C^{\mathcal{K},\psi}(\phi_g) = \min\left\{C^{\mathcal{K},\psi}(i) : \psi_i = \psi_0\right\} \geqslant d.$$

This follows from the fact that the minimum is achieved either at $i = 0$ or at $i = 0^d 1m$ with $m \notin \mathcal{M}'$. Neither have a representation using a function with a ψ-index without the $0^d 1$ prefix.

Choosing $d \geqslant r + \max\left\{C^{\mathcal{K},\phi}(\psi_0), c_g\right\} + c_f + 2q$ ensures that for both cases, we have $C^{\mathcal{K},\psi}(\phi_g) \geqslant C^{\mathcal{K},\phi}(\phi_g) + r + c_f + 2q$. While $C^{\mathcal{K}}(\psi_0)$ depends on the choice of d, we have $C^{\mathcal{K}}(\psi_0) \leqslant C^{\mathcal{K}}(d) + C^{\mathcal{K}}(D)$, up to a constant, which is in $O(\log d)$, so we can choose d to satisfy the inequality.

Finally, we can show the result:

$$C_{\text{comp}}^{\{g'\},\psi}(x) - C_{\text{comp}}^{\{f'\},\psi}(x)$$

$$= C^{\mathcal{K},\psi}(\phi_g) + \min\{|y| : \phi_g(y) = x\}$$

$$- C^{\mathcal{K},\psi}(\phi_f) - \min\{|y| : \phi_f(y) = x\}$$

$$\geqslant C^{\mathcal{K},\phi}(\phi_g) + r + c_f + 2q + \min\{|y| : \phi_g(y) = x\}$$

$$- C^{\mathcal{K},\phi}(\phi_f) - c_f - \min\{|y| : \phi_f(y) = x\}$$

$$= C_{\text{comp}}^{\{g\},\phi}(x) - C_{\text{comp}}^{\{f\},\phi}(x) + r + 2q \geqslant r. \qquad \square$$

Theorems 4.1 and 4.2 follow as corollaries. For Theorem 4.1:

Proof. Let D be a prefix function as in Lemma 4.3 that returns the index of u for the argument ϵ and ∞ for any other argument. That is, $\mathcal{M}' = \{u\}$. This construction satisfies the conditions 1 and 2 from Lemma 4.3. Invoking it, we find that there exists an acceptable numbering ψ for which $S^{\mathcal{M},\mathcal{N},\psi,k}(x) = S^{\mathcal{M}',\mathcal{N},\phi,k}(x) + c$. Since $\mathcal{M}'$ contains only a single model, $S^{\mathcal{M}',\mathcal{N},\phi,c}(x)$ is constant. $\qquad \square$

And for Theorem 4.2:

Proof. Let x be any string. Given a description of x, we construct some index i such that $\phi_i(\epsilon) = x$ (a singleton for x). Thus, $C^{\mathcal{K},\psi}(\phi_i) \leqslant C^{\mathcal{K},\psi}(x)$ up to a constant. Likewise, given ϕ we can produce x, so that $|C^{\mathcal{K},\phi}(\phi_i) - C^{\mathcal{K},\phi}(x)| \leqslant c$ for some constant c.

We now define a computable function D by $D(\bar{i}y) = j$ where $\phi_j(\epsilon) = \phi_i(y)$ and $i \in \mathcal{K}$, and let $\mathcal{M}'$ be its range. We will show that the two conditions of Lemma 4.3 hold for the prefix function D.

(a) Let $f \in \mathcal{M}'$ with $\phi_f(\epsilon) = x$. Then $\min\{|p| : \psi_{D(p)} = \phi_f\} = \min\{|\bar{i}q| : \phi_i(q) = x\} = C^{\mathcal{K}}(x) \leqslant C^{\mathcal{K}}(\phi_f) + c$. (b) On the one hand $C_{\text{comp}}^{\mathcal{M}',\psi}(x) \leqslant C^{\mathcal{K}}(\phi_f) + |\epsilon| \leqslant C^{\mathcal{K}}(x) + c$. On the other hand, the witness to $C_{\text{comp}}^{\mathcal{M},\psi}(x)$ is an effective description of x, so $C^{\mathcal{K}}(x)$ is at most a constant larger.

Now, by Lemma 4.3 there is a numbering ψ such that we have

$S^{\mathcal{M},\mathcal{N},\psi,k}(x) = S^{\mathcal{M}',\mathcal{N},\psi,k}(x) + c$. We observed that $|C^{\mathcal{K}}(\phi_i) - C^{\mathcal{K}}(x)| \leqslant c_0$ for all singletons, so $S^{\mathcal{M}',\mathcal{N},\psi,k}(x) \geqslant C^{\mathcal{K},\psi}(x) - c_0$. This proves the theorem. $\qquad\square$

Thus, in this balanced sophistication, there is no invariance: all information can be seen as structure, or as noise, depending on the numbering. To avoid these issues, existing proposals upset the balance to exclude or penalize the universal models, and possibly the singleton models.

4.4.1 Overfitting

We will now review the treatments in the literature that show overfitting. The first is the structure function, proposed by Kolmogorov, most likely the first attempt at separating structure from noise in an objective manner. Kolmogorov defined the following function, using the finite sets $\mathcal{F}$ as models:

$$h_x(\alpha) = \min\left\{\log |F| : x \in F, C^{\mathcal{K}}(F) \leqslant \alpha\right\}$$

and suggested that the smallest set for which $C^{\mathcal{K}}(F) + \log |F| \leqslant C^{\mathcal{K}}(x) + c$ holds for some pre-chosen constant c, can be seen as capturing all the structure in x [31]. This is equivalent to the sophistication $S^{\mathcal{F},\mathcal{K},\psi,c}(x)$. Theorem 4.2 shows there are numberings for which this sophistication is always equal to $C^{\mathcal{K}}(x)$. Thus, either this is true for all numberings, or this sophistication is not invariant.

In [37] the structure function is extended to an *algorithmic sufficient statistic*. This is, again, essentially the witness to the sophistication $S^{\mathcal{F},\mathcal{K},\psi,c}(x)$. A probabilistic version is also introduced, which uses the model class $\mathcal{P}$, which indexes the set of functions that compute computable probability semimeasures up to a multiplicative constant error, yielding $S^{\mathcal{P},\mathcal{K},\psi,c}(x)$. For both, Lemma 4.2 gives us a numbering such that the singleton is always the minimal sufficient statistic.

It may be argued that the slack parameter c in the sophistication, which determines the allowed gap between a candidate representation and the complexity, should depend on the numbering, but this dependence has not been mentioned in the literature and there is no obvious method to choose this constant for a given numbering.

In traditional statistics, overfitting is often addressed by a penalty on complex models. As we have seen, a strong penalty, such as the one imposed by an inefficient prefix encoding of the model, will cause underfitting. A more subtle approach is to allow descriptions that are not self-delimiting. The gap between the smallest self-delimiting description and the smallest non self-delimiting description grows without bound [63, Section 4.5.5], so that some information ends up in the noise, since placing all information in the model results in a self-delimiting, and thus non-optimal description. This eliminates the singletons as viable candidates. This approach is taken by Vitányi [91] and by Adriaans [8]. Such measures reduce the overfitting problem, but they only increase the tendency to underfit. We also pay the price that the models can no longer be equated with probability measures, weakening the link to traditional statistics.

4.4.2 Underfitting

Universal models are a widely acknowledged problem for sophistication, and most proposals avoid them by limiting the allowed models to exclude them. It is known that there are strings x for which $S^{\mathcal{F},\mathcal{K},\psi,c}(x)$, $S^{\mathcal{J},\psi,c}(x)$ and $S^{\mathcal{J},\mathcal{K},\psi,c}(x)$ are close to $|x|$ (up to a logarithmic term). Proofs can be found in [37], [13] and [91] respectively. These are the *absolutely non-stochastic strings* [83]. The existence of these strings is independent of the numbering.

However, the problem of the singletons remains. Only one model class eliminates both the singletons and the universal model: $\mathcal{J}$. The only proposal we are aware of that uses an efficient model representation *and* excludes the universal models *and* excludes the singletons is: $S^{\mathcal{J},\mathcal{K},\psi,c}$, from [91]. While this avoids our proofs of boundedness, there is no evidence that $S^{\mathcal{J},\mathcal{K},\psi,c}$ is actually invariant.

While high sophistication strings exist for $S^{\mathcal{J},\mathcal{K},\psi,c}$, they may not conform to sophistication's motivating intuition. To show this, we use the concept of depth:

Definition 4.3 (Depth[20, 11]). Let U be some universal Turing machine, so that $U(\bar{i}y) = \phi_i(y)$. Let U^t be a simulation of this machine, which is allowed to run for at most t steps, and returns 0 if it has not yet finished at that point. Let $C_t^{\mathcal{M}}(x) = \min\{|\bar{\imath}y| : U^t(\bar{\imath}y) = x, \phi_i \in \mathcal{M}\}$. The c-*depth* is

$$d^{\mathcal{M},c}(x) = \min\left\{t : C_t^{\mathcal{M}}(x) - C^{\mathcal{M}}(x) \leqslant c\right\} .$$

Deep strings are those that can only be optimally compressed with a great investment of time. As we saw in the last chapter, it is exceedingly unlikely that a deep string is sampled from a shallow distribution [20].

Theorem 4.3. Let $A(n)$ be the single-argument Ackermann function and c_d some constant. For all k, there is a numbering ψ such that for all strings x with depth $d^{c,c_d}(x) \leqslant A(C^c(x))$ the sophistication $S^{\mathcal{T},\mathcal{K},\psi,k}(x)$ is bounded.

Proof. Let $U(\bar{\imath}y)$ be a universal Turing machine, and let $U^A(\bar{\imath}y)$ be a simulation of that machine which outputs 0 if the number of steps taken exceeds $A(|\bar{\imath}y|)$. Let u be the index of the function U^A in the standard enumeration.

Let $D(\epsilon) = u$. We can instantiate Lemma 4.3 with D, $\mathcal{M}' = \{u\}$ and $X = \{x : d^{c,c_d}(x) \leqslant A(C^c(x))\}$. This tells us that there exists a numbering ψ for which $S^{\mathcal{T},\mathcal{K},\psi,k}(x) = S^{\mathcal{M}',\mathcal{K},\psi,k}(x) + |\bar{0}| \leqslant c$ for all $x \in X$. $\qquad\square$

This shows that while high-sophistication strings exist, they do not behave as expected. Consider a string that is typical for a shallow model, say some elaborate probabilistic automaton. Under $S^{\mathcal{T},\mathcal{K},\psi,c}$, no matter how high the complexity of the automaton, the sophistication is bounded. We could encode the collected works of Shakespeare in its transition graph, and this information would be counted as noise. Any structure simple enough to be exploited within the time bound of the Ackermann function will not be seen as 'meaningful information'. Only structure so deep that it would take beyond the lifetime of the universe to decompress would count towards sophistication. In the remainder we will refer to strings x with $d^{c,c}(x) \leqslant A(C^c(x))$ as *shallow* strings. Note that any string whose shortest program can be run in any time bound represented by a primitive recursive function is shallow.

The relation between $S(x)$ and $d(x)$ is also investigated in [12], where it is shown that within a logarithmic error term on the sophistication and the slack, they are identical. Our point is not the similarity between the two, but that for all practical strings, the sophistication is bounded. This contradicts the intuition that sophistication measures structure, as it seems to suggest that all strings we can possibly hope to understand or generate contain no structure, save a constant amount. The alternative is that under other numberings these strings *do* have structure, but then the sophistication is not invariant.

As for the strings with high sophistication, they have the property that they can be compressed far better with partial functions than with total: they are non-typical for the model class $\mathcal{T}$. This suggests that the 'non-stochastic' property of strings with high sophistication [83, 90] says more about depth and totality than it does about structure and noise.

4.4.3 Other variants

By moving away from the idea of two-part coding, the mechanics of lemma 4.3 can be avoided. In [68], the *naive sophistication* is introduced. We will define a generic version, parametrized by model class. Let $C_{\psi_i}(x) = \min\{|y| : \psi_i(y) = x\}$. Then we define the *naive sophistication* as:

$$S^{\mathcal{M},\psi,c}_{\text{naive}}(x) = \min\left\{C^{\mathcal{K},\psi}(\psi_i) : C_{\psi_i}(x) - C^{\mathcal{K},\psi}(x \mid i) \leqslant c, i \in \mathcal{M}\right\}.$$

The condition now is not that the two-part code length is minimal, but that the *randomness deficiency* $C_{\psi_i}(x) - C^{\mathcal{K},\psi}(x \mid i)$ is less than a constant. $S^{\mathcal{F},\psi,c}_{\text{naive}}(x)$ corresponds to the version in [68]. The switch to the randomness deficiency avoids Theorem 4.2, but we end up with the same problem as in Theorem 4.3: for shallow strings $S^{\mathcal{T},\psi,c}_{\text{naive}}$ is defined by the model U^A, and thus bounded.

We cannot show that $S^{\mathcal{F},\psi,c}_{\text{naive}}(x)$ is bounded for shallow strings, but this is only a consequence of the use of sets, not of the switch to randomness deficiency as a condition. *Any* set sophistication is necessarily lower-bounded by the function $\text{set}(x) = \min\{C^{\mathcal{K}}(F) : x \in F\}$ and if this function were bounded, it would suggest that a finite amount of finite sets contained all strings.

Theorem 4.4. Let ψ be *any* acceptable numbering. Then for all shallow x and large enough c, $S_{\text{naive}}^{\mathcal{J},\mathcal{K},\psi,c}(x)$ is bounded and for come constant c_F, we have:

$$\text{set}(x) \leqslant S_{\text{naive}}^{\mathcal{J},\mathcal{K},\psi,c}(x) \leqslant C^{\mathcal{K},\psi}(C^{\mathcal{K},\psi}(x)) + c_F.$$

Proof. Let ψ be any acceptable numbering. Let the Turing machine U be defined as $U(\bar{\imath}y) = \psi_i(y)$ if $i \in \mathcal{K}$, and $U(\bar{\imath}y) = \infty$ otherwise. Let U^A be derived from U as in section in Section 4.4.2 and let ϕ_u compute U^A.

For the first part we have $C_{\psi_u}(x) - C^{\mathcal{K},\psi}(x \mid u) \leqslant c_0$ for some c_0, thus for large enough c, $S_{\text{naive}}^{\mathcal{J},\psi,c}(x) \leqslant C^K(\psi_u)$. For the second part, let $C^{\mathcal{K},\psi}(x) = k$ and $F_k^A = \{x \mid \exists p : U^A(p) = x, |p| = k\}$. $|F_k^A| \leqslant |\{p : |p| = k\}|$, so that $\log |F_k^A| \leqslant k$, which gives us $\log |F_k^A| - C^{K,\psi}(x \mid F_k^A) \leqslant c_1$. Thus, for large enough c, $S_{\text{naive}}^{\mathcal{J},\psi,c}(x) \leqslant C^K(F_k^A)$. From a description of k, we can compute F_k^A with a finite program, so that $C^{\mathcal{K},\psi}(F_k^A) \leqslant C^{\mathcal{K},\psi}(k) + c_F$, which completes the proof. $\square$

Note that the constant c only needs to be large enough to ensure that $C^{\mathcal{K},\psi}(x) - C^{\mathcal{K},\psi}(x \mid u) \leqslant c$ and $C^{\mathcal{K},\psi}(x) - C^{\mathcal{K},\psi}(x \mid F_k^A) \leqslant c$. Since u and F_k^A are generally of no value in computing x, c is likely very small.

Another approach is the *coarse sophistication* [13], defined in [68] as:

$$S_{\text{coarse}}^{\mathcal{M},\mathcal{N},\psi}(x) = \min_c \left\{ S^{\mathcal{M},\mathcal{N},\psi,c}(x) + c \right\}.$$

Again, this variant avoids the pitfalls of Theorem 4.2. If there are candidates that are as good as the singletons but with smaller size by more than a constant, the constant penalty c will eventually be much less than the gain for the simpler witness, and the singletons will not determine the coarse sophistication. The coarse sophistication is within a logarithmic term of the busy beaver depth [13]. As with the naive sophistication, we can show that for shallow strings, the total function version is bounded, and the set version grows very slowly:

Theorem 4.5. Let ψ be *any* acceptable numbering. For all shallow x, $S_{coarse}^{\mathcal{T},\mathcal{K},\psi}(x)$ is bounded and there is a constant c_F such that:

$$set(x) \leqslant s_{coarse}^{\mathcal{F},\mathcal{K},\psi}(x) \leqslant 2C^{\mathcal{K},\psi}(C^{\mathcal{K},\psi}(x)) + c_F.$$

Proof. Let ψ be any acceptable numbering and define U^A, ϕ_u and F_k^A as in the proof of Theorem 4.4. For the first part, we know that for some constant c_0, $C_{comp}^{\{u\},\psi}(x) \leqslant C_{comp}^{\mathcal{K},\psi}(x) + c_0$ so that $S^{\mathcal{T},\mathcal{K},\psi,c_0}(x) \leqslant C^{\mathcal{K},\psi}(\phi_u)$, thus $S_{coarse}^{\mathcal{T},\mathcal{K},\psi}(x) \leqslant C^{\mathcal{K},\psi}(\phi_u) + c_0$. For the second part, we know that for some c_1, $C^{\mathcal{K},\psi}(F_k^A) + \log|F_k^A| \leqslant C^{\mathcal{K},\psi}(k) + C^{\mathcal{K},\psi}(x) + c_1$, so that $S^{\mathcal{F},\mathcal{K},\psi,C^{\mathcal{K}}(k)+c_1}(x) \leqslant C^{\mathcal{K},\psi}(F_k^A)$. Thus, $S_{coarse}^{\mathcal{F},\mathcal{K},\psi}(x) \leqslant C^{\mathcal{K},\psi}(F_k^A) + C^{\mathcal{K},\psi}(k) + c_1 = 2C^{\mathcal{K},\psi}(C^{\mathcal{K},\psi}(x)) + c_F.$ □

In [89], Vereshchagin proposes a *strongly algorithmic sufficient statistic*. Where the regular algorithmic sufficient statistic F from [37] has $C^{\mathcal{K}}(F \mid x)$ constant, the strong variant imposes the stronger requirement that $C^{\mathcal{T}}(F \mid x)$ is also constant. This reduces the problems of underfitting discussed in this section, but since $C^{\mathcal{T}}(\{x\} \mid x)$ is bounded, by Theorem 4.2, overfitting remains a problem: there exist numberings under which the singletons are the only candidates.

Finally, *effective complexity* [40], proposed by Gell-Man and Lloyd, was formulated from the perspective of physics, but fits the mold of sophistication. The model class consists of all computable probability distributions on finite sets. The complexity of the model is measured by its Kolmogorov complexity, avoiding the problems of Section 4.3. Theorem 4.2, however, still applies to effective complexity. Unlike other sophistication measures, it is not the candidate with the smallest model which is chosen, but the one which reproduces the data within the shortest time. Thus, if there are multiple candidates, this approach would likely favor the singletons. In [41], the authors abandon this strategy, and note that the choice from the set of candidates is a subjective one, which depends on context, which is in line with the view we express in the next section.

4.5   Discussion and conclusion

We have criticized existing measures of sophistication and shown technical problems with all of them. But that does not in itself mean that it should be impossible to come up with a sound measure. The common intuition, starting with the structure function, appears to be that the crucial property is whether a string is typical for a model, and that this typicality can be tested: another random choice from that model should select a string with the same structure. This idea is bold, but not unreasonable. Nevertheless, we offer the opinion that such a clean-cut separation *cannot* be made to work. We provide two arguments.

For the first argument, we take a generative perspective. We can generate data from a model $\phi_i, i \in \mathcal{K}$, by feeding it random bits until it produces an output. We will call the resulting probability distribution p_i. Call a sophistication *consistent* if, for sufficiently large data, it reflects the complexity of the source of the data. Now, let $\phi_u(\bar{\imath}y) = \phi_i(y)$ and sample from p_u. Then the initial bits will determine the prefix encoded index $\bar{\imath}$ of the function ϕ_i that ϕ_u will subsequently emulate, and the remaining bits are used as inputs to ϕ_i. We now ask, what should be the sophistication of the resulting data?

Certainly, if we have to judge based only on the data, we cannot exclude the possibility that the data was sampled from p_u: after all, it was. Yet, neither can we deny that it may have came from p_i, as again, it did. Eliminating the universal models does not solve this problem: the same argument holds if ϕ_u indexes, for instance, only those models computable by finite automata. *Any* model that dominates a set of other models creates this kind of ambiguity.

This shows clearly the limits of the single sample setting: with a second sample, the distinction would be easily made. The probability that a second sample from ϕ_u chooses ϕ_i again is negligable. If we get another sample that is likely to have come from ϕ_i, we have evidence to dismiss ϕ_u as a model.

Consider the following metaphor. We are given a a bitmap image of the painting *Impression of a Sunrise*. There are many good models for this string, from very generic to very specific. Sophistication suggests that we can choose one of these as the objective, intrinsic model of the data. The universal model says that it is 'some compressible, finite object'. Another might say that it is 'an image'. Even more specific would be 'a painting', 'a Monet', or specifically 'the painting *Impression of a Sunrise*'. A sound sophistication should be able to select one of these as the proper representation of structure in the data, and disqualify the others as over- or underfitting. But how should we be able to say that the data is intrinsically more of a painting than an image? More of a Monet than a painting? Intuitively, such distinctions require further assumptions, or a second sample from the same distribution.

The second reason we doubt sophistication is more technical. Consider the set of all possible two-part representations of x. When the numbering is changed, the codelength of the model part of all these representations will change. This is illustrated in the second diagram in Figure 4.1. The invariance theorem expresses that this change is limited by a constant term. However, even this small shift can push some representations out of the acceptable region (indicated by the dashed line), and pull others in. This may lead to a different representation determining the sophistication, one whose *total* codelength is close to what it was before, but whose *model* codelength can be anywhere between 0 and $C^e(x)$. If such jumps can occur, the sophistication is not invariant. And while we cannot *prove* in general that such jumps can always occur, there seems to be no reason to believe that they do not. Indeed, in [12] it is shown that logarithmic changes in the slack parameter can already cause these effects.

So we take a skeptical view of sophistication. Note that part of the theory is fine: there is nothing wrong with evaluating models for the data by comparing their two-part code lengths. In fact, the randomness deficiency $-\log p_i(x) - C^{\mathcal{K}}(x \mid i)$ has a direct statistical interpretation as a measure of counterevidence—under p_i, the probability of a randomness deficiency above k is less than 2^{-k} [23, Lemma 6]. In the Monet example above, this will allow us to disqualify the model expressing that the data is actually, say, a recording of jazz music.

But fundamental problems arise as soon as a hard cut-off is introduced on how far we are allowed to deviate from the minimum determined by the Kolmogorov complexity. In our opinion, a lot of measures taken in the literature, such as restricting the model class or introducing model penalties, complicate the method and make problems harder to analyze, without actually addressing the fundamental issue. This is dangerous: if such ad-hoc fixes result in a theory that is hard to prove either wrong or right, it creates an artificial dead end for a valuable area of research. When the hard cut-off on candidates is avoided, how-ever, all such measures are no longer necessary. What remains is an elegant theory that can be used to sift through all possi-ble models, disproving most while retaining a select number of interesting candidates for our further consideration.

In the next chapter, we take this approach to attack a prac-tical problem: finding patterns in graphs. Instead of attempt-ing to confirm the patterns as true—that is, performing model selection—we will compute several bounds on the Kolmogorov complexity and use the no-hypercompression inequality to reject models.

5 · COMPRESSION AS A MEASURE OF NETWORK MOTIF RELEVANCE

The material in this chapter is adapted from the paper Compression as a measure of network motif relevance, **P. Bloem**, *S. de Rooij, This paper is currently under submission at PLOS ONE.*

In our perspective, every statistician has a only a single sample. The fact that some may treat theirs as a series of independent draws is just an assumption about the source of that single sample. The data can be cut into chunks that are in some sense similar, and that similarity allows them to reconstruct the source of the data.

The benefit of this view is clear when we realize that other statisticians are not so lucky. For instance, those faced with graphs: a social network, a citation network or the webgraph. These are densely interconnected objects. Following friendship links on Facebook, we can trace a path between any two random subscribers through only 3.74 intermediaries on average [16]. Everybody is close to everybody else. This is one of the features that makes such networks so useful for things like information transfer. But it raises a question: if everybody is close to everybody else, what constitutes a "neighbourhood"? How do we slice up the network into similar pieces? We know that social networks, at least contain clusters of friends, but finding them is no small task.

A promising approach is that of *network motifs*: simply look for small subgraphs that recur frequently. These may point to communities, or to "functional units" of the network, performing the same task in different contexts. We investigate network motifs and show that the correspondence between descriptions and probabilities can be very valuable in the analysis of graphs.

Since the last chapter has shown that unless our model class is relatively limited, consistent model selection is likely a hopeless business, we will take a different tack. Instead of attempting to select the true model, we will use the no-hypercompression in-equality to *reject* models. This will not allow us to find "true" patterns, necessarily, but it will provide us with patterns that *might* be true. A list of candidates on which a domain expert can build. In short we will use the principles developed so far to perform *exploratory analysis*.

5.1   Network Motifs

Network motifs [67] provide an intuitive way to analyze graph structure. They are small, frequently occurring subgraphs. To be able to conclude that such frequent subgraphs really represent meaningful aspects of the data, we must first show that they are not simply a product of chance. That is, any subgraph may simply be a frequent subgraph in *any* random graph: a subgraph is only a *motif* if its frequency is *higher than expected*.

This expectation is defined in reference to a *null-model*: a probability distribution over graphs. We determine what the expected frequency of the subgraph is under the null-model, and if the observed frequency is substantially higher than this expectation, the subgraph is a motif. If the frequency is lower than expected, the subgraph is called an *anti-motif*.

The choice of null-model is an important aspect of the analysis. Consider the case explored in [24], where the data is directed and acyclic, as in the case of a citation graph. If the null model allows graph cycles, then any subgraph containing a cycle will be an *anti-motif*. Such motifs show only that the data is acyclic, and obscure any deeper structure. A model that produces random acyclic graphs will fit the data better, and will allow us to explore deeper structure. This shows the role of the null-model: the better we model the *known* structure in the data, the better we can expose the *unknown* structure.

However, there is usually no efficient way to compute the expected frequency of a subgraph under a null model. The most common approach is to generate a large number of random graphs, say 1000, from the null-model and compare the frequencies of the subgraph found in this sample to its frequency in the data [67]. This means that any resources invested in extracting the motifs from the data must be invested again 1000 times to find out which subgraphs are motifs.

We introduce an alternative method that does not require us to repeat the motif search on samples from the null model. We use two probability distributions on graphs: the null model $p^{null}(G)$, and a distribution $p^{motif}(G)$ under which graphs with one or more frequent subgraphs have high probability. If $p^{motif}(G)$ is larger than $p^{null}(G)$, the subgraph is a motif. Section 5.2 explains the principle and its theoretical justification.

To design p^{motif}, we make use of the Minimum Description Length (MDL) Principle [79, 47]. It can be shown that any description method L, a *code*, corresponds to a probability distribution p_L in such a way that a graph G with a short description under L will have a high probability under p_L. This correspondence is detailed in the preliminaries. Thus, we only need to design a code that exploits recurring subgraphs to give us a probability distribution that assigns graphs with recurring subgraphs higher probability. In brief, we accomplish this by describing the motif only once, and referring back to this description wherever the motif occurs. Since we do not need to describe the motif explicitly for every occurrence, graphs with a high frequency of a certain motif will have a short description length, and thus a high probability. The code is described in Section 5.3.

Our method has several advantages:

- The search for motifs only needs to be run once: on the data G. To compare the result against the null-model, we only need to know $p^{null}(G)$.
- The number of motif instances found does not need to be an accurate estimate of the number present in the graph. The only aim is to find a sufficiently large set of non-overlapping instances in the data, to prove that the subgraph is a motif. This allows faster and simpler search algorithms to be used.
- Given sufficiently strong evidence, a single test can be used to eliminate multiple null models. This is explained in Section 5.2.

We perform several experiments to validate these claims. First, we create random graphs with a number of occurrences of a specific subgraph inserted. We then show that our method can identify the subgraphs very precisely, even if only a small number were added. Secondly, we illustrate the behavior of the method on two directed, and two undirected graphs, using three different null models. Finally, to show what is possible with fast null models, we run the method on a dataset of a million nodes and 13 million links. This analysis was run in just under 6 hours in a single-threaded implementation, showing the scalability of the method.

All software and data used in this chapter is available under the MIT License.[1]

5.1.1 Related work

Motif analysis has been applied in many domains, such as the study of biological networks [94], the problem of community detection in social networks [7] and the investigation of neural networks [84]. Motif extraction is a form of *subgraph mining*. However, while general subgraph mining tends to focus on finding frequent subgraphs, motif extraction focuses more on the problem of finding *meaningful* subgraphs, usually with the help of significance tests. Our method facilitates the computation of the significance test, and can be combined with any subgraph mining algorithm.

The idea of the network motif was first introduced under that name in [67]. In that paper, a computationally expensive, comprehensive search for motifs was used. Later, in [52], a simple sampling algorithm was introduced which is able find the most frequent motifs of many graphs with as little as 500 samples. However, as noted in [93], it is highly biased.

A different solution to the problem of repeating the motif search on samples from the null model is provided in [93]: there, a faster and more correct sampling algorithm is provided, together with a technique to compute the subgraph frequencies indirectly from a single search on the data. However, this technique is restricted to the use of a specific null-model, and then only with a particular sampling method. As noted in the introduction, the restriction to one null-model is a serious drawback, and as noted in [42], this particular sampling method lacks strong guarantees

---

[1]https://github.com/Data2Semantics/nodes/wiki/Motifs

on mixing time. [77] provides a good overview of other algorithms available for motif analysis.

The idea that compression can be used as a heuristic for subgraph discovery was also used in the SUBDUE algorithm by Cook and Holder [30]. We introduce a different compression method, connect it to the framework of motif analysis, and make the statistical implications precise.

In this work, all candidate-motifs are induced subgraphs. This is not common to all motif analysis; in some settings the instances of the motif are allowed to have additional internal links that are not part of the motif [25]. While our method could be adapted to find such motifs, we will not discuss such adaptations here.

5.1.2 Preliminaries: graphs and codes

Graphs   A graph G of size n is a tuple (N, L) containing a set of *nodes* N and a set of *links* L. For convenience in defining probability distributions on graphs, N is always the set of the first n natural numbers. L contains pairs of elements from N. Let N_G be the nodeset of G and L_G be its linkset. For the dimensions of the graph we use the functions $n(G) = |N_G|$ and $m(G) = |L_G|$. If a graph G is *directed*, the pairs in L_G are ordered, if it is *undirected*, they are unordered. A *multigraph* has the same definition as a graph, but with L_G a multiset, i.e. the same link can occur more than once.

A *simple graph* is a graph where no link connects a node to itself. There are many types of graphs and tailoring a method to all of them is a laborious task. Here, we limit ourselves to datasets that are simple graphs. This is usually the most complex setting, so that we can trust that a method that works for simple graphs is easily translated to other settings.

Two graphs G and H are *isomorphic* if there exists a bijection $f : N_G \rightarrow N_H$ on the nodes of G such that two nodes a and b are adjacent in G if and only if $f(a)$ and $f(b)$ are adjacent in H. If two graphs G and H are isomorphic, we say that they belong to the same isomorphism class [G].

The distinction between G and [G] is important. Often, G is given with the nodes in arbitrary order and we are actually only interested in the properties shared by all graphs in [G]. However, such analyses on $[G]$ can prove to be very expensive. For this reason, almost all literature on complex networks, analyzes graphs rather than isomorphism classes. Sometimes, the result is the same in both cases. For instance, let p_a and p_b be two graph models that are both uniform within every isomorphism class, i.e. $\forall H \in [G] : p(H) = p(G)$. Then, the relative magnitude of $p_a(G)$ and $p_b(G)$ is the same as that of $p_a([G])$ and $p_b([G])$. There are other cases, however, where the analysis on G must be seen as an approximation to the desired analysis on $[G]$.

Codes   In previous chapters, we built the idea of a description method on top of Turing machines. In this chapter, it is more efficient to construct description methods directly from a correspondence to probability distributions, forgetting about Turing machines for the time being.

Let $\mathbb{B}$, again, be the set of all finite-length binary strings. We use $|b|$ to represent the length of $b \in \mathbb{B}$. Let $\log(x) = \log_2(x)$. A *code* for a set of graphs $\mathcal{G}$ is an injective function $f : \mathcal{G} \to \mathbb{B}$. It is *self-delimiting* if no code word is the prefix of another. We will denote a *codelength function* with the letter L, i.e. $L(G) = |f(G)|$. It is common practice to compute L directly, without explicitly computing the codewords. In fact, we will adopt the convention of referring to L itself as a code.

A well known result in information theory is the association between codes and probability distributions, implied by the *Kraft inequality*: for each probability distribution p^* on $\mathcal{G}$, there exists a self-delimiting code L^* such that for all $G \in \mathcal{G}$: $-\log p^*(G) \leqslant L^*(G) < -\log p^*(G) + 1$. Inversely, for each self-delimiting code L^* for $\mathcal{G}$, there exists a probability distribution p^* such that for all $G \in \mathcal{G}$: $p^*(G) = 2^{-L^*(G)}$. For proofs, see [47, Section 3.2.1] or [32, Theorem 5.2.1]. To give an intuition, note that we can easily transform a code L^* into a sampling algorithm for p^* by feeding the decoding function random bits until it produces an output. To transform a probability distribution to a code, techniques like *arithmetic coding* [81] can be used. Lemma 3.1 also provides evidence for this correspondence.

As explained in [47, page 96], the discrepancy between $-\log p^*(G)$ and $L^*(G)$ can be safely ignored and we may *iden-*

tify codes with probability distributions. Thus we allow L(G) to take non-integer values.

When we need to encode a single choice from a finite set S of options, we will often use the code with length log |S|, corresponding to a uniform probability on S.

5.2 Model selection by codelength

The association between codes and probability distributions is particularly useful in the design of graph models: many structural properties can easily be exploited to encode a graph efficiently. Consider an undirected graph G containing a large clique: all nodes in some subset $N_C \subseteq N_G$ are connected to one another directly. We can describe the graph by first describing N_C, and then describing G in a canonical manner. Since every node in N_C is connected to every other node in the clique, we can omit these links from the second part of our description, shortening the total description length, if N_C is large enough. By the correspondence mentioned in the preliminaries, this gives us not just a code L^{clique} with short codelengths for graphs with large cliques, but also a probability distribution p^{clique} with high probabilities for graphs with large cliques.

Of course, there is no guarantee that of all the distributions with a bias towards large cliques, p^{clique} matches the source of our data. Luckily, it does not need to. The presence of the clique disproves the hypothesis that the data came from the null-model, so long as we can show that our clique-based model encodes the data more efficiently.

We return to the *no-hypercompression inequality*: under the hypothesis that the null-model was the source of the data, we can show that the probability that any other model compresses the data better by k bits or more, decays exponentially in k. More precisely, let $p^{null}(x)$ be any probability distribution, with $L^{null}(x) = -\log p^{null}(x)$ and let L(x) be any code, then we have:

$$p^{null}\left(L^{null}(x) - L(x) \geqslant k\right) \leqslant 2^{-k}.$$

Thus, under the null-model, the probability that L^{clique} will compress the data better than the null-model by 10 bits or more is less than one in one-thousand. For twenty bits, we get one in a million, for thirty bits, one in a billion, and so on. So while a low codelength under L^{clique} does not prove that the clique-code is the true model, it does allow us to comfortably reject the null model.

We can interpret this procedure as a significance test: the difference in compression D between the null model and the alternative model is a statistic [47, Example 14.2]. The no-hypercompression inequality gives us a bound on the probability $p^{\text{null}}(D \geqslant k)$. To reject the null-model with significance level α, we must find some code on the set of all graphs and show that it compresses the data better than the null-model by k bits, with $2^{-k} \leqslant \alpha$. Any code will do, so long as it was chosen before seeing the data.

Note that D is also the logarithm of the likelihood ratio between the null model and L, so we can see this as a likelihood ratio test. We can also interpret the difference in codelength between two models p_a and p_b as the logarithm of the *Bayes factor* $p_a(x)/p_b(x)$ [47, Section 14.2.3].

Now, while our test only *rejects* the null-model, and does not confirm anything, we would like to make sure that it was the pattern we are interested in (e.g. the clique) that allowed us to reject the null model, and not some other aspect of the alternative model. To ensure this, we aim to have the alternative model exploit only the pattern, and nothing else. We use the null model for everything but the pattern. For instance, in the example above, the clique model must store the graph minus the links of the clique. If we use the null model for this, we know that the only change between the null model and the alternative is the use of the clique, so that must be what made the difference.

A final benefit of this method is that we can reject multiple null models with a single test. In many situations we will have a function $B(G)$ that lower bounds any code in some set $\mathcal{L}$. If our alternative model provides a codelength below $B(G) - k_\alpha$ with k_α the number of bits required for our chosen α, we can reject all of $\mathcal{L}$.

As an example, Let $\mathcal{G}_n$ be the set of all undirected graphs of size n. We can define a uniform code on such graphs:

$$L_n^{\text{uniform}}(G) = \log |\mathcal{G}_n| \text{ (for any } G \in \mathcal{G}_n).$$

This code captures the idea that the size of the graph is the only informative statistic: given the size, all graphs are equally likely. This is a good null-model to test the assumption that the graph contains no significant structure, save for its size. However, it is *parametrized*. It is currently not a code on *all* graphs, just those of size n. To turn it into a code that can represent all graphs, we need to encode the parameter n as well, with some code over the natural numbers

$$L^{\text{complete}}(G) = L^{\mathbb{N}}(n(G)) + L_{n(G)}^{\text{uniform}}(G).$$

This is called *two-part coding*, we encode the parameters of a model first, and then the data given the parameters. For some parametrized model L_θ, we can choose any code for θ to make it complete. We will call the set of all such complete codes the *two-part codes on L_θ*.

Which two-part code we choose is arbitrary. We may be able to reject the uniform code for one choice of $L^{\mathbb{N}}$, or several, but how can we prove that L^{complete} will be rejected whatever $L^{\mathbb{N}}$ we choose? Instead of choosing an arbitrary code for the size, we can instead use the *bound* $B(G) = L_{n(G)}^{\text{uniform}}(G)$ as our null model. This is not a code, but it *is* a lower bound for any two-part code on L_n^{uniform}. If $L^{\text{clique}}(G)$ is shorter than $B(G)$, it is also shorter than $L^{\text{complete}}(G)$ for any choice of $L^{\mathbb{N}}$.[2]

Contrast this with the traditional approach, where we would define a statistic on G, like the size of the largest clique, and compare the observed value of the statistic with the expectation under the null model. In this case the models would have to be rejected with separate tests. If a large clique is unlikely in a sample from p_n^{uniform}, we have no guarantee that it will also be unlikely in a sample from p^{complete}.

---

[2]In probabilistic terms, the code on the parameter corresponds to a prior on the parameter. The two-part codes correspond to maximum likelihood posterior probabilities: $p(\hat{\theta})p(x \mid \hat{\theta})$. Our bound corresponds to the maximal likelihood of the data: $p(x \mid \hat{\theta})$. This shows us that the bound applies not only to the two-part codes, but also to the full mixture: $\sum_\theta p(x \mid \theta)p(\theta) \leqslant \sum_\theta p(x \mid \hat{\theta})p(\theta) = p(x \mid \hat{\theta})$

Note that when we store the rest of the graph within L^{clique} we *cannot* use $B(G)$ in place of $L^{\text{complete}}(G)$. We want a *conservative* hypothesis test: the probability of rejecting a true null model may be lower than α but never higher. By this principle, bounds chosen in place of a model should always decrease D. The code corresponding to the null-model must always be lowerbounded, and the code for the alternative model must always be upperbounded. Thus when we re-use the null model inside the alternative model, we must always use a complete code.

5.3 Encoding with motifs

Let $S = \langle S_1, \ldots, S_k \rangle$ be an ordered set of nodes from N_G. The *induced subgraph* $I(S, G)$ is a graph G' with k nodes, containing a link (i, j) if and only if G has a link (S_i, S_j). That is, the induced subgraph extracts all links existing between members of S.

Assume that we are given a graph G, a potential motif G', and a list $M^{\text{raw}} = \langle M_1, \ldots, M_k \rangle$ of instances of G' in G. That is, each sequence $M \in M^{\text{raw}}$ consists of nodes in N_G, such that the induced subgraph $I(M, G)$ is equal to G'. Sequences in M^{raw} may overlap, i.e. two instances may share one or more nodes. We are also provided with a generic graph code $L^{\text{base}}(G)$ on the simple graphs.

The basic principle behind our code is illustrated in Fig. 5.1: we want to store the motif only once, remove as many instances of the motif from the data as we can, and replace them with references to the stored motif. The two graphs combined contain enough information to recover the data, but we have only had to describe the motif once. Algorithm 1 on page 97 describes the exact process.

The first thing we need for this scheme is a subset M of M^{raw} such that the instances contained within it do not overlap: ie for each M_a and M_b in M, we have $M_a \cap M_b = \emptyset$. Selecting the subset that would give us optimal compression is NP-Hard (as the set cover problem is reducible to it), so we must make do with an approximation. As we will see later, the most important factor is the number of links an instance has to nodes outside the instance. We call this the *exdegree*.[3] In order to find a subset

---

[3] Unlike the in- and outdegree the exdegree is not a property of a node, but

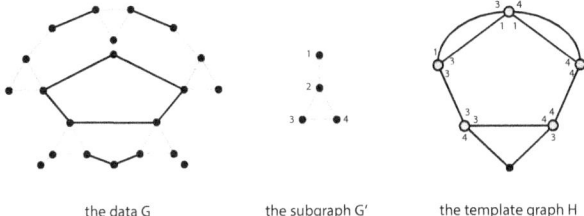

the data G   the subgraph G'   the template graph H

Figure 5.1: An illustration of the motif code. We store G' once, and remove its instances from G, replacing them with a single, special node. The links to special nodes are annotated with 'rewiring' information, which tells us how to rewire the subgraph back into H. Storing only H and G' is enough to reconstruct the data.

of instances with low exdegree, we first sort $\mathcal{M}^{\mathrm{raw}}$ by exdegree in ascending order. We then remove the first M, add it to our subset $\mathcal{M}$ and remove all other instances that overlap with it. We continue removing the first remaining instance until $\mathcal{M}^{\mathrm{raw}}$ is empty.

In the following, we will often need to store a sequence of integers. We will store all such sequences using the code corresponding to a *Dirichlet-Multinomial* (DM) distribution. Let S be a sequence of length k of elements in alphabet Σ. Conceptually, the DM distribution models the following sampling process: we sample a probability vector p on $[0, |\Sigma|]$ from a Dirichlet distribution with parameter vector α, and then sample k symbols from the categorical distribution represented by p. The probability mass function corresponding to this process can be expressed as:

$$p_\alpha^{\mathrm{DirM}}(S \mid k, \Sigma) = \prod_{i \in [1,k]} \mathrm{DirM}_\alpha(S_i \mid S_{1:i-1})$$

$$p_\alpha^{\mathrm{DirM}}(S_i \mid S', k, \Sigma) = \frac{f(S_i, S') + \alpha_i}{|S'| + \sum_i \alpha_i}$$

where $f(x, X)$ denotes the frequency of x in X. We use $\alpha_i = 1/2$

of a subgraph.

for all i. Let $L^{\text{DirM}}_{k,\Sigma}(S) = -\log p^{\text{DirM}}(S \mid k, \Sigma)$. Note that this code is parametrized with k and Σ. If these cannot be deduced from earlier parts of the code, they need to be encoded separately. Often, we will have $\Sigma = [0, n_{\max}]$, and we only need to encode $n_{\max}$.

We also require a self delimiting code to represent natural numbers. We will use the code corresponding to the probability distribution $p^{\mathbb{N}}(n) = 1/(n(n+1))$, and denote it $L^{\mathbb{N}}(n)$.

We can now describe how each part of the graph is encoded:

subgraph First, we store the subgraph G' using $L^{\text{base}}(G')$ bits.

template We then create the *template graph* H by removing the nodes of each instance $M \in \mathcal{M}$, except for the first, which becomes a specially marked node, called an *instance node*. The internal links of M—those incident to two nodes both in M—are removed from the graph. Any link connecting a node outside of M to a node inside of M is kept, and rewired to the instance node.

instance nodes L^{base} does not record which nodes of H are instance nodes, so we must record this separately. There are $\binom{n(G)}{|\mathcal{M}|}$ possibilities, so we can encode this information in $\log \binom{n(g)}{|\mathcal{M}|}$ bits.

rewiring For each side of a link in H incident to an instance node, we need to know which node in the motif it originally connected to. Let there be some agreed-upon order in which to enumerate the links of any given graph. Given this order, we only need to encode the sequence W of integers $w_i \in [1, \ldots, n(G')]$. We do so using the DM model described above. The maximum symbol and length of W can be deduced from parts already encoded. Note that this code is invariant to the ordering of W, so the particulars of the canonical node ordering do not need to be specified.

multiple edges Since L^{base} can only encode simple graphs, we cannot use it to store H directly, since collapsing the instances into single nodes may have created multiple edges. In this case we remove all multiple edges and encode them separately. We assume a canonical ordering over the links and record for each link incident to an instance node, how many copies of it were removed. This gives us a sequence

R of natural numbers $R_i \in [0, r_{max}]$ which we store by first recording the maximum value in $L^N(\max(R))$ bits, and then recording R with the DM model.

insertions Finally, while H and G' give us enough information to recover a graph isomorphic to G, we cannot yet reconstruct where each node of a motif instance belongs in the node ordering of G. Note that the first node in the instance became the instance node, so we only need to record where to insert the rest of the nodes of the motif. This means that we perform $|\mathcal{M}|(n(G') - 1)$ such insertions. Each insertion requires $\log(t + 1)$ bits to describe, where t is the size of the graph before the insertion. Let H be the template graph and G the complete graph, then we require $\sum_{t=n(H)}^{n(G)-1} \log(t+1) = \log(n(G)!) - \log(n(H)!)$ bits to record the correct insertions.

search Since our code accepts any list of motif instances, we are free to take the list $\mathcal{M}$ and prune it further, before passing it to the motif code, effectively discounting instances of the motif. This can often improve compression, as storing the rewiring information for instances with high exdegrees may cost more than we gain from removing them from the graph. We will sort $\mathcal{M}$ by exdegree and search for the value c for which compressing the graph with only the first c elements of $\mathcal{M}$ gives the lowest codelength.

The codelength L^{motif} as a function of c is roughly unimodal, which means that a ternary search should give us a good value of c while reducing the number of times we have to compute the full codelength. We use a *Fibonacci search* [53], an elegant variation on ternary search, requiring only one sample per recursion. Note that c is not a parameter of the model, so we do not need to store it separately.

implementation The **template** part of the code can be time and memory intensive for large graphs, as it involves creating a copy of the data. For any given L^{base}, we can create a specific implementation which computes the codelength required for storing the template graph without constructing H explicitly. This will speed up the computation of the code at the expense of creating a new implementation for each new null-model. We use such specific implementations for our three null-models.

5.4   Null models

We will define three null-models. For each model we follow the same pattern, we first describe a parametrized model (which does not represent a code on all graphs). We then use this to derive a bound as described in the second section, so that we can reject a set of null models, and finally we describe how to turn the parametrized model into a complete model to store graphs within the motif code.

Specifically, let $L_\theta^{name}(G)$ be a parametrized model with parameter θ. Let $\hat{\theta}(G)$ be the value of θ that minimizes $L_\theta^{name}(G)$ (the maximum likelihood parameter). From this we derive a bound $B^{name}(G)$ from this—usually using $B^{name}(G) = L_{\hat{\theta}(G)}^{name}(G)$—which we will use in place of a null-model. Finally, we create the complete model by two-part coding: $L^{name}(G) = L^\theta(\hat{\theta}(G)) + L_{\hat{\theta}(G)}^{name}(G)$.

5.4.1   The Erdős-Renyi model

The Erdős-Renyi (ER) model is probably the best known probability distribution on graphs [76, 43]. It takes a number of nodes n and a number of links m as parameters, and assigns equal probability to all graphs with these attributes, and zero probability to all others. This gives us

$$L_{n,m}^{ER}(G) = \log \binom{(n^2 - n)/2}{m}$$

for undirected graphs, and

$$L_{n,m}^{ER}(G) = \log \binom{n^2 - n}{m}$$

for directed graphs. We use the bound $B^{ER}(G) = L_{n(G),m(G)}^{ER}(G)$.

For a complete code on simple graphs, we encode n with $L^\mathbb{N}$. For m we know that the value is at most $m_{max} = (n^2 - n)/2$ in the undirected case, and at most $m_{max} = n^2 - n$ in the directed case, and we can encode such a value in $\log m_{max}$ bits:

$$L^{ER}(G) = L^\mathbb{N}(n(G)) + \log m_{max} + L_{n(G),m(G)}^{ER}(G).$$

5.4.2 The degree-sequence model

The most common null-model in motif analysis is the *degree-sequence model* (also known as the *configuration model* [73]). For undirected graphs, we define the degree sequence of graph G as the sequence $D(G)$ of length $n(G)$ such that D_i is the number of links incident to node node i in G. For directed graphs, the degree sequence is a pair of such sequences $D(G) = (D^{in}, D^{out})$, such that D_i^{in} is the number of incoming links of node i, and D_i^{out} is the number of outgoing links.

The parametrized model $L_D^{DS}(G)$ The degree-sequence model $L_D^{DS}(G)$ takes a degree sequence D as a parameter and assigns equal probability to all graphs with that degree sequence. Assuming that G matches the degree sequence, we have $L_D^{DS}(G) = \log|\mathcal{G}_D|$ where $\mathcal{G}_D$ is the set of simple graphs with degree sequence D. There is no known efficient way to compute this value for either directed or undirected graphs, but various estimation procedures exist. We use an importance sampling algorithm discovered independently by [22] and [42].[4] This algorithm is guaranteed to produce any graph matching D with some nonzero probability. Crucially, the algorithm does not backtrack or reject candidates, which means that if we multiply the probability of each random choice made in sampling, we get the probability of the sample under our sampling procedure. That is, the algorithm produces, along with a sample $G \in \mathcal{G}_D$, the probability $q_D(G)$ of the algorithm producing G. While the samples are not uniform, we do have:

$$E\left[\frac{1}{q_D(G)}\right] = |\mathcal{G}_D| \qquad (5.1)$$

where G is a random variable representing a sample from the algorithm. Thus, we can sample a number of graphs and take the mean of their inverse probability under q_D to estimate $p_D^{DS}(G)$. This is a form of *importance sampling*.

This approach was taken in [22]: the sample mean $1/n \sum_i 1/q_D(G_i)$ was used as an estimator for the expectation in (5.1). However, as shown in [42], the distribution of $1/q_D(G)$ tends to be very close to log-normal. This means that the sample mean will converge *very* slowly to the correct value. Specifically, the standard deviation of this estimate after n samples is $\frac{1}{\sqrt{n}}\sqrt{(e^{\sigma^2}-1)e^{2\mu+\sigma^2}}$, which for a distribution with $\mu = 200$ and $\sigma = 10$, leads to a standard deviation of approximately $\frac{1}{\sqrt{n}}e^{300}$.

For this reason, we use the *maximum-likelihood estimator* for the log-normal distribution instead. Let $Q_i = 1/q_D(G_i)$. We assume Q_i is log-normally distributed, so that $Y_i = \log Q_i$ is normally distributed. Let $\overline{Y} = n^{-1}\sum_i Y_i$ and $S_Y = 1/n\sum_i(\overline{Y}-Y_i)^2$; then the maximum-likelihood estimator of EQ is $\exp\left(\overline{Y}+\frac{1}{2}S_Y\right)$. Thus, the codelength under the degree sequence model can be estimated as $\left(\overline{Y}+\frac{1}{2}S_Y\right)\log_2(e)$.

Unfortunately, even with the highly optimized implementations described in [42] and [54] sampling can be slow for large graphs. Luckily, we are only interested in an estimate of the codelength accurate to around the level of single bits, which means that we only need to sample until we have a rough estimate of the order of magnitude of $|\mathcal{G}_D|$. For instance, if we accept a margin of error of only 15 bits (of the potentially 10^6 bits required to store a medium-sized graph), we can underestimate the number of graphs by 4 orders of magnitude and still end up within the margin. All we need is a reliable confidence interval for our estimate, so that we can choose a suitably conservative bound. Our method of obtaining such a confidence interval is described in the appendix. In all cases, we use a one-sided confidence interval: when computing the codelength under the null model, we use a lower bound for the true value, and when computing the codelength for the motif code, we use an upper bound. Thus, the difference in codelength is a lower bound for the true value.

The bound $B^{DS}(G)$ To get a bound for all two-part codes on L_D^{DS}, we could use $B'(G) = L_{D(G)}^{DS}(G)$. Beating such a bound would tell us that no property of the degree sequence could explain the motif we had found. Unfortunately, the degree sequence forms a large part of the code, and a lot of evidence is

---

[4]Specifically, our implementation uses the algorithms described in [42] and [54]. However the non-uniform sampling from the candidate set, discussed in [22, p10, step 5] is crucial to achieving a low variance in the sampling distribution, and thus a fast convergence.

required to compress better than $B'(G)$ with a complete code.

Instead, we make the assumption that the degrees are sampled independently from a single distribution $p^{deg}(n)$ on the the natural numbers. This corresponds to a code $\sum_{D_i \in D} L^{deg}(D_i)$ on the entire degree sequence. Let $f(s, D)$ be the frequency of symbol s in sequence D. It can be shown that

$$B^{deg}(D) = -\sum_{D_i \in D} \log \frac{f(D_i, D)}{|D|}$$

is a lower bound for any such code on the degree sequence. This gives us the bound $B^{DS}(G) = B^{deg}(D(G))) + L^{DS}_{D(G)}(G)$. For directed graphs, we use $B^{DS}(G) = B^{deg}(D^{in}(G))) + B^{deg}(D^{out}(G))) + L^{DS}_{D(G)}(G)$.

The complete model $L^{DS}(G)$ For the alternative model we need a complete code. First, we store $n(G)$ with $L^{\mathbb{N}}$. We then store the maximum degree and encode the degree sequence with the DM model. For undirected graphs we get:

$$L^{DS}(G) = L^{\mathbb{N}}(n(G)) + L^{\mathbb{N}}(\max(D)) + L^{DirM}_{n(G),\max(D)}(D) + L^{DS}_{D(G)}(G)$$

and for directed graphs

$$L^{DS}(G) = L^{\mathbb{N}}(n(G))$$

$$+ L^{\mathbb{N}}(\max(D^{in})) + L^{DirM}_{n(G),\max(D^{in})}(D^{in})$$

$$+ L^{\mathbb{N}}(\max(D^{out})) + L^{DirM}_{n(G),\max(D^{out})}(D^{out}) + L^{DS}_{D(G)}(G).$$

Note that in the computation of L^{motif} with L^{DS} as a base model, we estimate $|\mathcal{G}_D|$ for both the template graph and the motif. It is important to combine the confidence intervals over these two estimates carefully, so that we end up with a correct confidence interval over the total codelength. This is discussed in the supporting materials. For L^{motif}, we compute a one-sided confidence interval to get an *upper*bound, so that with 95% confidence we are *over*estimating the size of the motif code.

5.4.3 The edgelist model

While estimating $|\mathcal{G}_D|$ can be costly, we can compute an upper bound efficiently. Assume that we have a directed graph G with n nodes, m links and a pair of degree sequences $D = (D^{in}, D^{out})$. To describe G, we might write down the links as a pair (F, T) of sequences of nodes: with F_i the node from which link i originates, and T_i the node to which it points. Let S_D be the set of all pairs of such sequences satisfying D. We have $\binom{m}{D^{in}_1, \ldots, D^{in}_n}$ possibilities for the first sequence, and $\binom{m}{D^{out}_1, \ldots, D^{out}_n}$ for the second. This gives us $|S_D| = \binom{m}{D^{in}_1, \ldots, D^{in}_n}\binom{m}{D^{out}_1, \ldots, D^{out}_n} = m!m! / \prod_{i=1}^{n} D^{in}_i! D^{out}_i!$. We have $|S_D| > |\mathcal{G}_D|$ for two reasons. First, many of the graphs represented by such a sequence pair contain multiple links and self-loops, which means they are not in $\mathcal{G}_D$. Second, the link order is arbitrary: we can interchange any two different links, and we would get a different pair of sequences, representing the same graph, so that for a graph with no multiple edges, there are $m!$ different sequence-pairs to represent them.

To refine this upper bound, let $S'_D \subset S_D$ be the set of sequence pairs representing simple graphs. As all links in such graphs are distinct, we have $|\mathcal{G}_D| = |S'_D|/m!$. Since $|S'_D| \leqslant |S_D|$, we have [5]

$$|\mathcal{G}_D| \leqslant \frac{m!}{\prod_{i=1}^{n} D^{in}_i! D^{out}_i!} .$$

In the undirected case, we can imagine a single, long list of nodes of length $2m$. We construct a graph from this by connecting the node at index i in this list to the node at index $m + i$ for all $i \in [1, m]$. In this list, node a should occur D_a times. We define S_D as the set of all lists such that the resulting graph satisfies D. There are $\binom{(2m)!}{D_1, \ldots, D_n}$ such lists. We now have an additional reason why $|S_D| > |\mathcal{G}_D|$: each pair of nodes describing a link can be swapped around to give us the exact same graph. This gives us:

$$|\mathcal{G}_D| \leqslant |S'_D|/(2^m m!) = \frac{(2m)!}{2^m m! \prod_{i=1}^{n} D_i!} .$$

In both cases, the fact that we have an upperbound gives us a

---

[5] This value was previously used in [21] as a precise value for the number of graphs with multiple edges. This is incorrect, as we can only divide by $m!$ if we know that no graphs have multiple edges.

code: while the code as described assigns some probability mass to non-simple graphs, we can easily assume that this is assigned instead to some null-element, since we are only interested in the codelengths and probabilities of simple graphs. This gives us the following parametrized code for directed graphs:

$$L_D^{EL}(G) = \log m! - \sum_{i=0}^{n} \log D_i^{in}! - \sum_{i=0}^{n} \log D_i^{out}!$$

where (D^{in}, D^{out}) are the degree sequences of G, and for the undirected case:

$$L_D^{EL}(G) = \log(2m)! - \log m! - m - \sum_{i=0}^{n} \log D_i!.$$

For the bound and the complete model, we follow the same strategy we used for the previous model. For undirected graphs:

$$B^{EL}(G) = B^{deg}(G) + L_{D(G)}^{EL}(G), \text{ and}$$

$$L^{EL}(G) = L^{\mathbb{N}}(n(G)) + L^{\mathbb{N}}(\max(D)) + L_{n(G),\max(D)}^{DirM}(D) + L_{D(G)}^{EL}(G).$$

And for directed graphs:

$$L^{EL}(G) = L^{\mathbb{N}}(n(G))$$

$$+ L^{\mathbb{N}}(\max(D^{in})) + L_{n(G),\max(D^{in})}^{DirM}(D^{in})$$

$$+ L^{\mathbb{N}}(\max(D^{out})) + L_{n(G),\max(D^{out})}^{DirM}(D^{out}) + L_{D(G)}^{EL}(G).$$

5.5   Experiments

To validate and illustrate our method, we will perform three experiments. First, we will construct a graph by injecting instances of a single motif into a random network. The method should recover only this motif as significant. Second, we will run the method on datasets from four different domains, and show the results for the most frequent subgraphs, using the three null-models we have described. Finally, to show the scalability of the model with fast null models, we will run the analysis on a large graph.

In all experiments we search for motifs by sampling, based on the method described in [52]. Note that we have no particular need for a sampling algorithm which provides an accurate approximation of the actual frequencies present in the graph, so long as it can provide us with a large selection of non-overlapping instances with low exdegree. For this reason we adapt the algorithm to improve its speed: we start with an empty selection of nodes N', and add a random node drawn uniformly from N_G. We then add to N' a random neighbour of a random member of N', and repeat this action until N' has the required size. We then extract and return $I(N', G)$. In the case of a directed graph, nodes reachable by incoming and outgoing links are both considered neighbours.

The size $n(G')$ of the subgraph is chosen before each sample from a uniform distribution over the interval $[n_{min}, n_{max}]$. Where n_{min} and n_{max} are parameters of the experiment.

We re-order the nodes of the extracted graph to a canonical ordering for its isomorphism class, using the Nauty algorithm [66]. We maintain a map from each subgraph in canonical form to a list of instances found for the subgraph. After sampling is completed, we end up with a set of potential motifs and a list of instances for each, to pass to the motif code described in Section 5.3.

In all experiments we report the log-factor:

$$B^{null}(G) - L^{motif}(G; G', \mathcal{M}, L^{null}).$$

That is, we use the bound in place of the null model, and the complete code of the same null model is passed to the motif code. If the log-factor is larger than 10 bits, we can interpret it, as described in the Section 5.3, as a successful significance test, allowing us to reject the null model at $\alpha = 0.001$. In all cases, a negative log-factor means that we do not have sufficient evidence to reject the null-model, but a different experiment might yet achieve a positive log-factor. This could be achieved by sampling more subgraphs, using a different algorithm to find motif instances or taking more samples from the degree-sequence estimator.

Note that we do not correct for multiple testing, since this is purely an exploratory analysis and such corrections would not affect the relative ordering of the motifs. Nevertheless, it is important to bear this in mind when interpreting the log-factors.

5.5.1 Recovering motifs in synthetic data

We use the following procedure to sample an undirected graph with 5000 nodes and 10000 links, containing n^i injected instances of a particular motif G' with n' nodes and m' links:

1. Let $n = 5000 - (n' - 1)n^i$ and $m = 10000 - m'n^i$ and sample a graph H from the uniform distribution over all graphs with n nodes and m links.

2. Label n^i random nodes, with degree 5 or less, as instance nodes.

3. Let p^{cat} be a categorical distribution on $\{1, \ldots, 5\}$, chosen randomly from the uniform distribution over all such distributions.

4. Label every connection between an instance node and a link with a random value from p^{cat}. Links incident to two instance nodes, will thus get *two* values.

5. Reconstruct the graph G from G' and H.

This is roughly similar to sampling from our motif code. In this graph, G' should be the only significant motif, with the exception of motifs that can be explained from the prevalence of G', i.e. subgraphs and supergraphs of G', or graphs that contain part of G'. However, these should have a markedly lower log-factor than G'. For our experiment, we will only extract subgraphs of size 5, to rule out the first two cases.

On this sampled graph, we run our motif analysis. We run the experiment multiple times, with $n^i = 0$, $n^i = 10$ and $n^i = 100$, using the same subgraph G' over all runs, but sampling a different H each time. For each value of n^i, we repeat the experiment 10 times. Per run we sample 5000 motifs. This value is chosen to show that even a very *low* sample size is sufficient to recover the motif. The null-model in all cases is the ER model, as that corresponds to the source of the data.

Fig. 5.2 shows the results for the 21 possible connected simple graphs of size 5. As expected, when we insert no subgraphs, the motif model cannot compress the graph better than the null model, for any motifs, since the source of the data *is* the null-model. There are motifs with very high frequencies (shown on the right), much higher than the frequencies of our motif, but these can be explained as a consequence of the null model and have a negative log-factor. We can also see that once we insert 100 instances of the motif, two other subgraphs become motifs: in both cases, these share a part of the inserted motif (a rectangle and a triangle). This is an important lesson for motif analysis: not every motif represents a meaningful result, some motifs may be a byproduct of other motifs.

5.5.2 Various datasets and null-models

Next, we show how our our approach operates on a selection of datasets across domains. We use the following datasets:

kingjames **(undirected,** $n = 1773, m = 9131$**)** Co-occurrences of nouns in the text of the King James bible [2, 1]. Nodes represent nouns (places and names) and links represent whether these occur together in one or more verses.

yeast **(undirected,** $n = 1528, m = 2844$**)** A network of the protein interactions in yeast, based on a literature review [75]. Nodes are proteins, and links are reported interactions between proteins. We removed 81 self-loops.

physicians **(directed,** $n = 241, m = 1098$**)** Nodes are physicians in Illinois [5, 28]. Links indicate that one physician turns to the other for advice.

citations **(directed,** $n = 1769, m = 4222$**)** The arXiv citation network in the category of theoretical astrophysics, as created for the 2003 KDD Cup [39]. To create a workable graph, we follow the procedure outlined in [24]: we include only papers published before 1994, remove citations to papers published after the citing paper, and select the largest connected component.

All datasets are simple (no multiple edges, no self-loops). In each case we take $5 \cdot 10^6$ samples with $n_{min} = 2$ and $n_{max} = 6$. We test the 100 motifs with the highest number of instances (after overlap removal), and report the log-factor for each null model. For the edgelist and ER models we use a Fibonacci search at full depth, for the degree-sequence model we restrict the search depth to 3. For the degree-sequence estimator, we use 40 samples and $\alpha = 0.05$ to determine our confidence interval. We use the same set of instances for each null-model.

Our first observation is that for the physicians dataset, there are no motifs under the degree-sequence null-model. We have found this to be a common property of many social networks. Whether this indicates that social networks are simpler, more random, or perhaps even well-modeled by the degree sequence model, requires further investigation. We may be tempted to draw the conclusion that directed networks contain fewer motifs in general for these null models, or that fewer motifs can be found in directed networks with this method, but the experiment in the next section shows that this is not the case.

In both the kingjames and the yeast graphs, many motifs contain cliques or near-cliques. This suggests that the data contains local communities of highly connected nodes which the null model cannot explain.

We also observe a degree of agreement between the degree sequence model and the edgelist model, suggesting that the edgelist model may be an acceptable proxy for the degree sequence model.

These analyses were run on a single machine with 8 Gigabyte of java heapspace[6] and 2 1.80 Ghz Intel Xeon processors (E5-2650L). For the kingjames dataset the time taken was, on average, 23 minutes and 34 seconds per motif. For the yeast dataset, 3 minutes and 8 seconds per motif. For the physicians dataset, 35 seconds per motif and for the citations dataset 9 minutes and 25 seconds per motif. In all these experiments the degree-sequence model took by far the most time. The next section shows the possibilities if this model is not considered.

The sampling from the degree-sequence estimator was done in parallel, taking advantage of the 16 cores available in total. All other code was single-threaded.

---

[6]This was the system default. The amount of memory used was not measured.

5.5.3 Large-scale motif extraction

In the experiments above, the computation of L_D^{DS} was by far the greatest bottleneck. In order to test the scalability of the method for null-models which can be computed efficiently, we omit the degree-sequence null model. This allows us to perform motif detection on much larger graphs. We use the hyperlink graph of the Dutch Wikipedia [6, 74] as a benchmark. This dataset contains all links that existed at some point between any two articles of the Dutch Wikipedia. We removed self loops and multiple edges, resulting in a network of 1 039 253 nodes and 13 485 902 links. We sampled $5 \cdot 10^6$ subgraphs of sizes from 2 to 6 nodes. We selected the 100 motifs with the greatest number of instances (after overlap removal) and computed their log-factor under the ER and edgelist models. The top 30 are shown in Fig. 5.7. We used the Fibonacci search at full depth for both models.

The analysis was executed using 3.5 gigabytes of Java heapspace, on a machine with 2 1.80 Ghz Intel Xeon processors (E5-2650L). It took 5 hours and 27 minutes to complete. Sampling of subgraphs took 9 minutes, overlap removal took 14 minutes and the compression analysis took, on average, 3 minutes and 2 seconds per motif (for 100 motifs analyzed).

Note that all code in this analysis was single-threaded, so the availability of multiple cores and multiple processors was *not* exploited. While it is a simple matter to run the computation of the log-factors in parallel, with a thread for each motif, a single-threaded run shows that the experiment can also be run on commodity hardware, in reasonable time.

5.6   Conclusion

We have introduced a new method of testing motif relevance, which allows motif analysis to be scaled up to graphs with millions of nodes, even on commodity hardware.

One observation from our experiments deserves further mention: in the first experiment, we saw that injection of one subgraph into a network caused other subgraphs to become motifs, i.e. their frequencies became statistically significant. This tells us that even *if* some motifs represent functional units of a network, as is often claimed (and contested) [67, 56], the fact that a subgraph occurs with statistically significant regularity cannot be taken as proof that it is a functional unit. Hypothesis testing allows one to make a binary decision, but that decision is always *about the null-model*. A low p-value should not be interpreted as evidence for the meaning of the subgraph. In fact, at this level of abstraction, the best any method can do is to offer sound *candidates* for functional units.

The proof that a particular motif actually corresponds to a meaningful unit can only be achieved in context: that is, a domain expert should evaluate the list of instances found for a particular motif, to see whether a large subset of them perform the same role in the network, or if not, what other reason can be found for the prevalence of the motif. In other words, motif analysis is necessarily an *exploratory* technique, and while a significance test provides a good heuristic to separate trivially frequent subgraphs from subgraphs which may represent important properties, it is ultimately just a heuristic. The only thing it *proves* is the incorrectness of the null-model.

The reader may note that while we have shown that our method is fast in principle, if we wish to use the degree-sequence model, we are still limited to medium-scale graphs, and long processing times. There are several potential ways around this problem. First, the codelength for the whole graph is not specific to motif analysis. It only needs to be computed once for each G, after which it can be re-used in any MDL or Bayesian analysis (for motifs, cliques, clustering, etc.) Second, while we have chosen to use the null model also inside the motif model, this is not the only approach. We could also use the edgelist model in the motif code: since the edgelist code upperbounds the degree-sequence code, any positive log-factor in this setting means we would also get a positive log-factor if we used the degree-sequence model to store the motif and the template graph. There may also be efficient lowerbounds for the degree-sequence model, which we can use in its place for the null model. [7]

Our current model does not allow the detection of anti-motifs.

---

[7] [18] provides a lowerbound, but it cannot be computed much faster than the estimators used in our experiments. It may, however, be easier to parallelize.

For that purpose, another model would be required; one which exploits the property that a subgraph has a *lower* frequency than expected to compress the data. In theory, this is certainly possible: any such non-randomness can be exploited for the purposes of compression. We leave this as a matter for future research.

Finally, we hope that our approach is illustrative of the general benefit of MDL techniques in the analysis of complex graphs. In conventional graph analysis a researcher often starts with a structural property that is observed in a graph, and then attempts to construct a process which generates graphs with that structural property. A case in point is the property of scale-freeness and the preferential attachment process that was introduced to explain it [9]. The Kraft inequality allows us instead to build models based on a description method for graphs. The trick then becomes to find a code that describes such graphs with the desired property efficiently, instead of finding a process that is likely to generate graphs with the desired property. For many properties, such as cliques, motifs or specific degree sequences, such a code readily suggests itself.

Algorithm 1 The motif code $L^{\text{motif}}(G; G', \mathcal{M}, L^{\text{base}})$. Note that the nodes of the graph are integers.

Given:

a graph G, a subgraph G',

a list $\mathcal{M}$ of instances of G' in G, a code L^{base} on the simple graphs.

$b_{\text{subgraph}} \leftarrow L^{\text{base}}(G')$ **subgraph**

 # replace each instance with a single node
$H \leftarrow \text{copy}(G)$, $W = []$ **template**
for each $M = \{m_1, \ldots m_{n(G')}\}$ **in** $\mathcal{M}'$:

    # We use m_1 (the m_1-th node in G) as the instance node

    for each link l between a node n_{out} not in M and a node m_j in M:

        if $j \neq 1$: add a link between n_{out} and m_j

        $W.\text{append}(j)$

    remove all nodes m_i except m_1, and all incident links

$b_{\text{rewiring}} \leftarrow L^{\text{DirM}}_{|W|,n(G')}(W)$ **rewiring**

#Remove multiple edges from H and record the duplicates in R
$R, H' \leftarrow \text{simple}(H)$
$b_{\text{template}} \leftarrow L^{\text{base}}(H')$
$b_{\text{multi-edges}} \leftarrow L^{\mathbb{N}}(\max(R)) + L^{\text{DirM}}_{|R|,\max(R)}(R)$ **multiple edges**

$b_{\text{instances}} \leftarrow \log \binom{n(G)}{|\mathcal{M}'|}$ **instance nodes**
$b_{\text{insertions}} \leftarrow \log(n(G)!) - \log(n(H)!)$ **insertions**

return $b_{\text{subgraph}} + b_{\text{template}} + b_{\text{rewiring}} + b_{\text{multi-edges}} + b_{\text{instances}} + b_{\text{insertions}}$

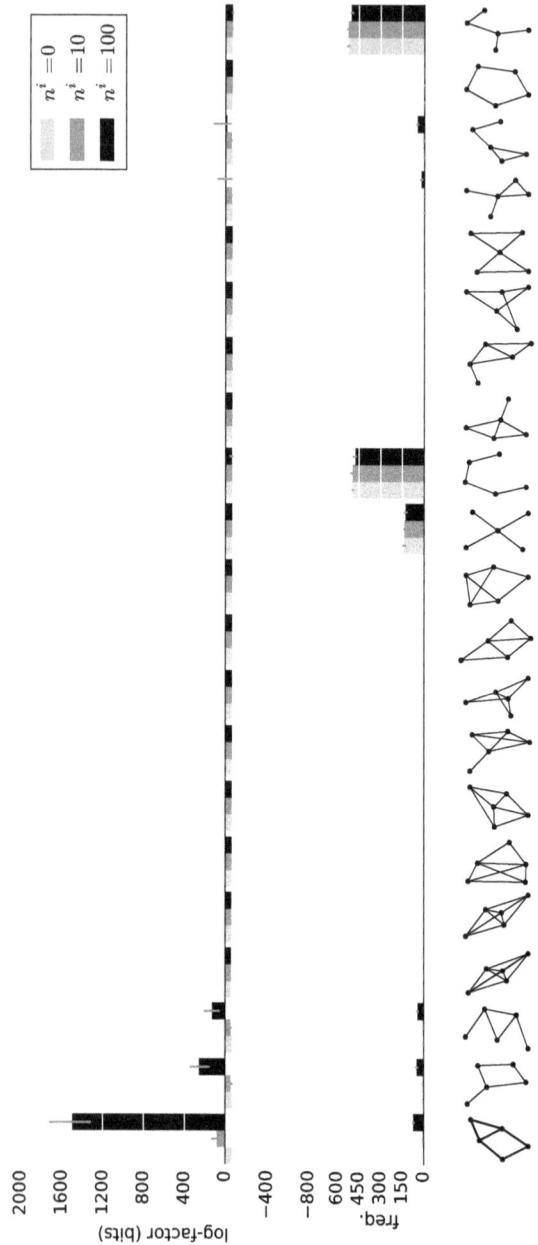

Figure 5.2: The results of the experiment on synthetic data. The bottom row shows all 21 simple connected graphs with 5 nodes (up to isomorphism). The middle row shows the number of non-overlapping instances found by the sampling algorithm for $n^i = 0$, $n^i = 10$ and $n^i = 100$ from left to right, for each motif. The bars show the average value over 10 randomly sampled graphs, with the same subgraph (shown on the far left) injected each time. The top row shows the difference between the codelength under the null model (the ER model) and under the motif code. The error bars represent the *range*, i.e. they are drawn from the smallest to the largest observation.

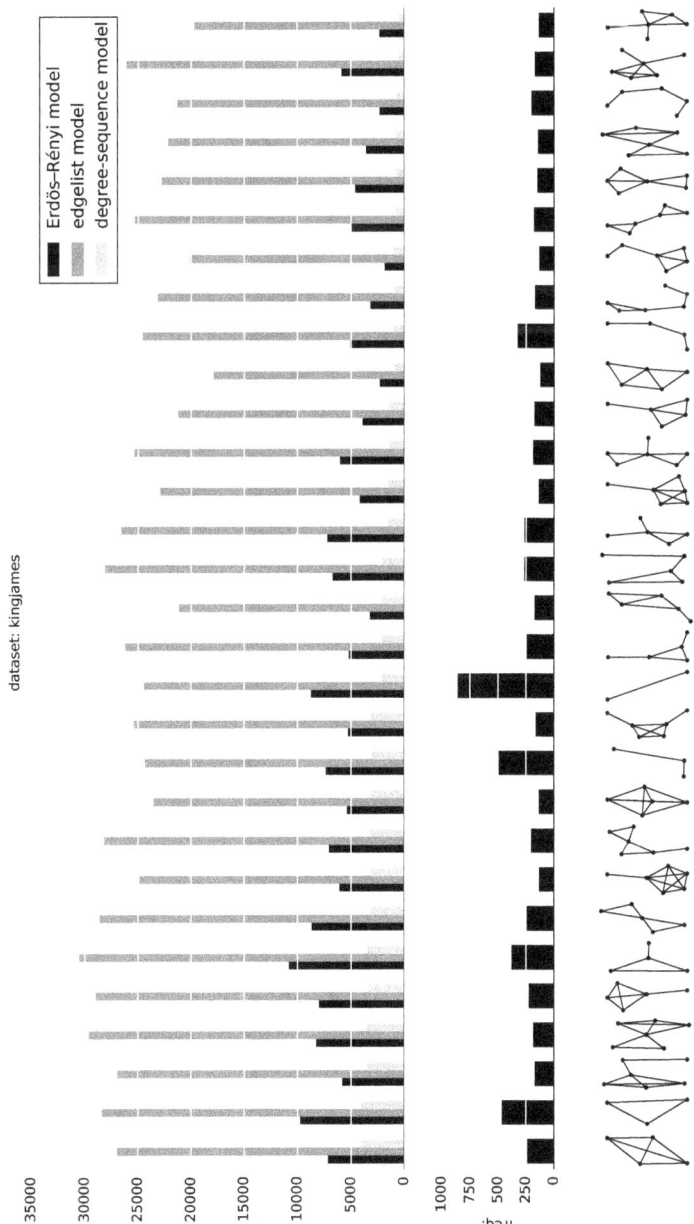

Figure 5.3: The results of the motif extraction on the kingjames network.

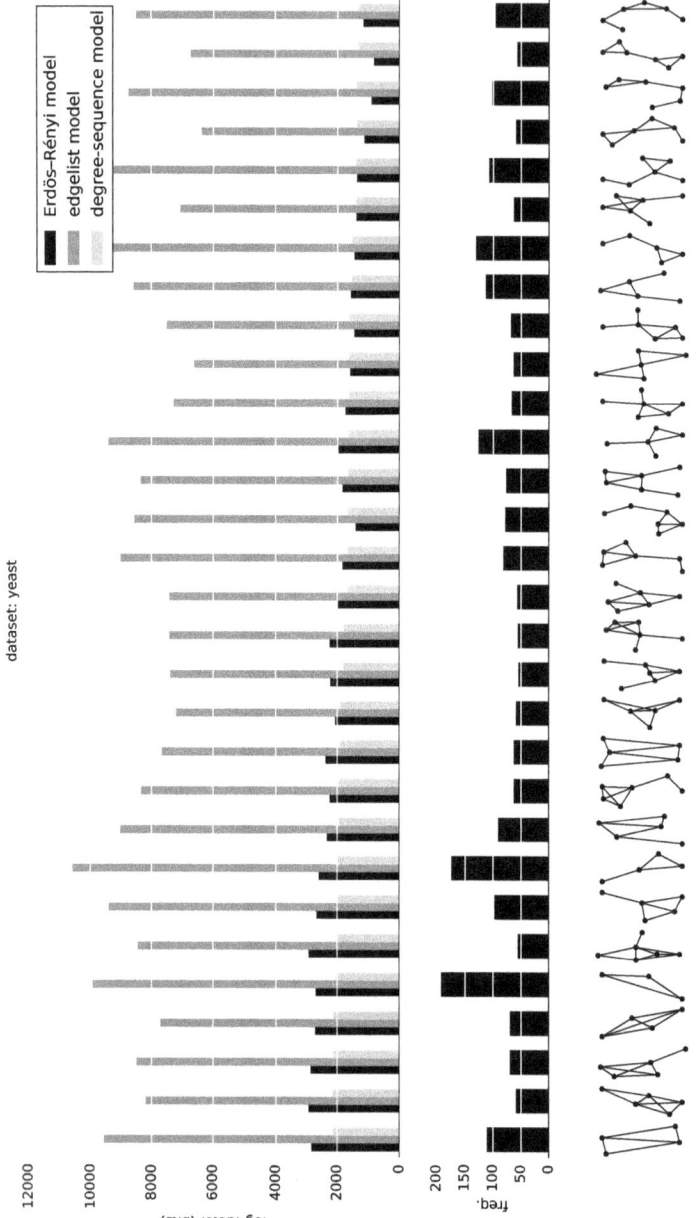

Figure 5.4: The results of the motif extraction on the yeast network.

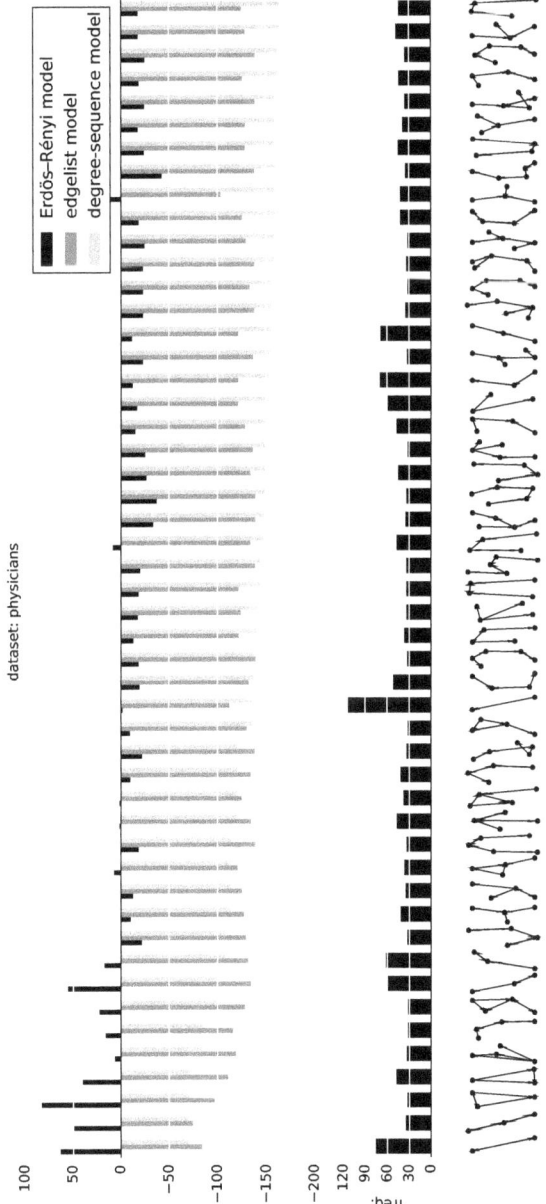

Figure 5.5: The results of the motif extraction on the physicians network.

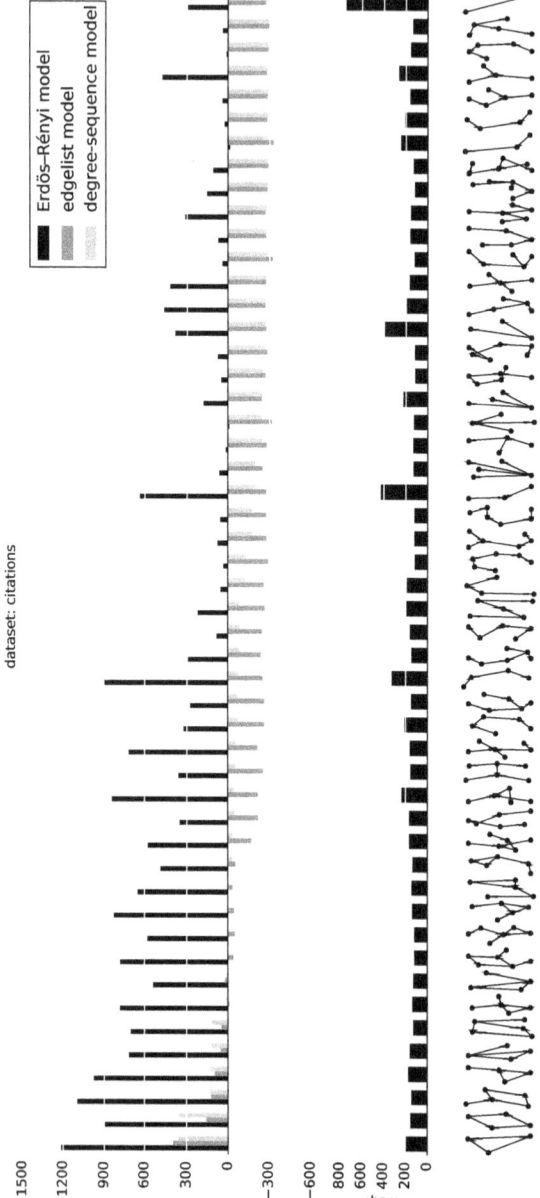

Figure 5.6: The results of the motif extraction on the citation network.

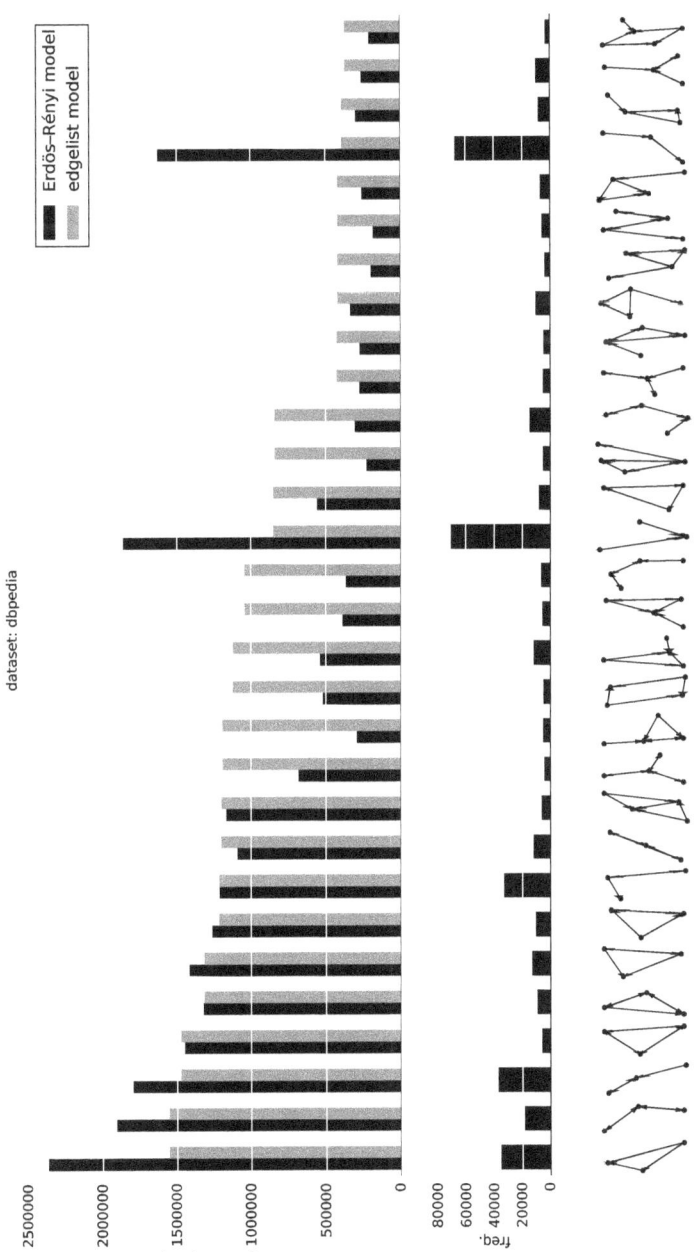

Figure 5.7: The results of the motif extraction on a large-scale network. We show the thirty motifs with the highest log-factor under the EL model.

6 · AN **EM** ALGORITHM FOR THE FRACTAL

INVERSE PROBLEM

The material in this chapter is based on the paper An EM-algorithm for the fractal inverse problem, **P. Bloem**, *S. de Rooij. To be submitted to* Physical Review Letters.

In the previous chapters, we saw that often, the easiest data to work with is data that consists of independent, identically distributed samples. Such data is made up of chunks of equal length, drawn from the same distribution independently. We can view this as a kind of *symmetry under translation*. Slide the bit string a fixed number of bits to the left or right, and the distribution over the part we're looking at remains invariant.

As we saw, this kind of symmetry is difficult to find in graphs. There is, however, another kind of symmetry in graph data: *symmetry under re-scaling*, also known as *self-similarity*: if we "zoom in" on the data in the correct way, looking at only a small part, this small part resembles the whole. This principle is commonly used implicitly: no researcher actually has access to the entire web-graph, for instance. They use a subgraph, extracted by crawling links on the web. Thus, they make the implicit assumption that conclusions drawn on the basis of this subgraph also hold for the whole. A subgraph, sometimes much smaller than the full graph, is assumed to be similar to the whole set.

This assumption has some interesting consequences. If the subgraph is like the whole in every respect, is a subgraph of the subgraph also like the subgraph, and thereby like the whole? Can the whole dataset be split in increasingly smaller sections that are all scaled down versions of the whole? If so, the result is a *fractal*.

Fractal models have been popular since the seventies, when mathematicians and physicists began to realize that they could explain jagged, irregular features like coastlines, clouds and trees. Objects that had proved increasingly difficult to analyze with classical Euclidean geometry. One problem that has always held

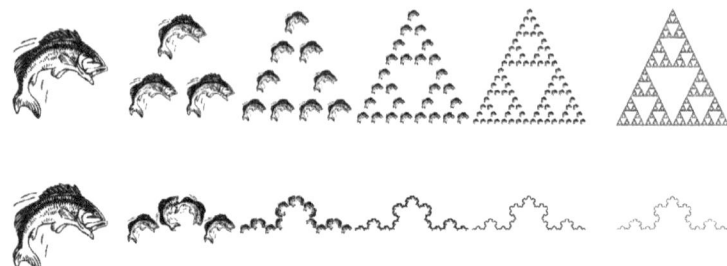

Figure 6.1: Two examples of fractals and their construction. In both cases, we start with an initial image, we take several transformed copies of this image and compose them into a new image. Iterating the process, the information in the original image disappears, and the fractal emerges. The figures on the right represent the limit of this process: the Sierpinski triangle and the Koch curve, respectively.

back fractal analysis, however, is the *inverse problem*: we know how to draw pictures of a given fractal model, but how to reconstruct the model given the picture?

In this chapter, we investigate the problem in the domain of iid numerical data: we consider the problem of fitting self-similar models to sets of points in an H-dimensional Euclidean space. We show that the inverse problem can be efficiently solved with Expectation-Maximization (EM) methods.

6.1 Fractals

Fractals are a class of mathematical objects, specifically sets or probability distributions. The word fractal is not precisely defined,[1] but the following properties are common to most examples. First, there is **self similarity**: a part is a scaled down copy of the whole. See for instance the Sierpinski triangle (Figure 6.1). It consists of three triangular shapes which are scaled down copies of itself. Second, most fractals have **infinitely fine structure**. 'Zooming in' reveals ever finer detail. In the case of the Sierpinski triangle, we will see the same shape recurring again and again, but other fractals like the Mandelbrot set reveal a great variety of shapes. Finally, there exist generalizations of the notion of dimension that can take on non-integer values. Most fractals will have such **non-integer dimensions**. The Sierpinski triangle, for example, has a *Hausdorff* dimension of approximately 1.58.

Since the name *fractal* was coined in the 1970s, fractals have increasingly been seen as a potential model for many natural phenomena. Mandelbrot put it as follows in *The Fractal Geometry of Nature* [64]:

> Clouds are not spheres, mountains are not cones, coastlines are not circles, and bark is not smooth, nor does lightning travel in a straight line.

Fractal geometry has been used in many fields, including physics [65], geology [26], biology [44] and economics [87].

One of the greatest problems with fractal analysis has always been the difficulty of finding a fractal model for a given set of data. It may be visually clear that a cloud or a coastline 'looks' fractal, and we may be able to determine that it has a non-integer dimension [86], but how do we get from a dataset to a model? How do we determine the parameters of the fractal-generating model that led to this particular image or dataset? This is called the *fractal inverse problem*.

Current approaches tend to rely on evolutionary algorithms [34, 29, 72]. Such algorithms are expensive, and it can be difficult to get them to converge to precise parameter values, even if the dataset itself is sampled from a fractal model. Other approaches are highly domain-specific, such as fractal image compression [48], which does not translate well to data of arbitrary dimensionality. Some interesting results have been also been derived from the method of moments [78] and sampling random transformations from the data [49], but apart from the evolutionary methods, we are not aware of any other generic, practical methods.

Here, we focus on the family of *Iterated Function Systems* [17, 51], a broad class of fractals capturing many well-known examples. Figure 6.1 shows the basic principle: we start with some initial image and apply a small number, K, of contracting transfor-

---

[1]Mandelbrot originally defined a fractal as a set whose topological dimension differs from its Hausdorff dimension, but later retracted this definition, stating that he preferred the word to be not precisely defined.

mations to it, resulting in a second image consisting of K scaled down copies of the initial image. We then apply the transformations again, to this second image, and so on. As we iterate this process, the information in the initial image is lost, and the image converges to a fractal. Which fractal emerges is entirely determined by the chosen transformations. By fixing a family of transformations F, we define a family of fractals, each fractal determined by a small number of transformations from F, its *components*.

This idea generalizes very naturally to probability distributions: we choose the components of the IFS, and instead of applying them to an initial image, we apply them to an initial *distribution*, combining these into a mixture of scaled-down copies of the original distribution. Thus, we can define a family of distributions on $\mathbb{R}^H$, where the components form the parameters of the model. Here, we choose *similitudes* as our family of transformations. A similitude, also known as a rigid transformation, consists of a translation, a rotation, and a uniform scaling. We choose similitudes because they offer a good trade-off between expressiveness, and ease of optimization. Most well-known IFSs can be described by a similitude-based IFS, and we can still solve the optimizations in our algorithm analytically. Extensions to other families of transformations are possible at the cost of using numeric or approximate optimization instead.

We can now frame the fractal inverse problem as a problem of statistical parameter estimation: for a set of points $x \in \mathbb{R}^H$, and a number of components K, find a set of K similitudes, such that the likelihood of X under the resulting IFS distribution is maximal. We use the Expectation-Maximization principle to construct our algorithm. Briefly, we cast the sequence of components that was responsible for a particular point as a *latent variable*. Optimizing both the latent variables and the parameters of the model together is intractable, but we can approximate by optimizing the latent variables given the parameters and vice versa. Starting with some initial model, we iterate this process, until we converge to a good model.

We show in Section 6.4 that our algorithm is able to reconstruct known fractals, such as the Sierpinski triangle and the

Koch curve. We also apply the algorithm to some datasets sampled from images, and some natural data of higher dimension, showing that, while there are no IFSs to perfectly capture these images, the self similarity in the model does often allow a better fit than a simple mixture of MVNs.

6.1.1 Preliminaries

Measures and transformations   Let $\{x_i\}_{i \in [1,N]}$, $x_i \in \mathbb{R}^H$ be our dataset. Let X be the $N \times H$ matrix with x_i as its columns. All vectors are column vectors.

Let 1^L be the length-L vector with 1 at all indices. Such vectors are useful tools in matrix notation: for instance, the sum of the elements of the $K \times L$ matrix M can be expressed as $1^{K^T} M 1^L$, while its marginal vectors are $(1^{K^T} M)^T$ and $M 1^L$. Since the correct length is often determined by context, we will omit the superscripts where possible. Let e^i be the vector with element i 1 and all others 0. This vector functions as a row- or column-selector, e.g. $Xe^i = x_i$.

Let $\nu(S)$ be a probability distribution with $S \subseteq \mathbb{R}^H$. Let $f_{t,A}(x) = Ax + t$ be an invertible affine transformation represented by a vector t and a matrix A. Then the transformation of ν by $f_{t,A}$ is defined by the relation $f_{t,A}(\nu)(S) = \nu(f_{t,A}^{-1}(S))$. Let $\nu(x)$ be the density function of the measure ν (with respect to the Lebesgue measure). Then the density function of the measure $f_{t,A}(\nu)$ is $f_{t,A}(\nu)(x) = |A^{-1}| \nu(f_{t,A}^{-1}(x))$.

A multivariate normal distribution (MVN) on $\mathbb{R}^H$ is determined by a mean $\mu \in \mathbb{R}_H$ and a covariance matrix $\Sigma \in \mathbb{R}^{H \times H}$. Its probability density function is:

$$\mathcal{N}(x; \mu, \Sigma) = (2\pi)^{-\frac{H}{2}} |\Sigma|^{-\frac{1}{2}} \exp\left[-\frac{1}{2} \left|\left|(x - \mu)^T \Sigma^{-1} (x - \mu)\right|\right|^2 \right]$$

We will call the MVN with $\mu = 0$, $\Sigma = I^H$ the *standard multivariate normal distribution*, or $\mathcal{N}_0$. A *spherical MVN* has $\Sigma = sI$ for some scalar s. Let x be a random variable with $x \sim \mathcal{N}(\mu, \Sigma)$, then $f_{t,A}(x) \sim \mathcal{N}(t + A\mu, A\Sigma A^T)$.

A *rotation matrix* R is a matrix with $RR^T = I$ and $|R| = 1$. A *similitude* (or rigid transformation) on $\mathbb{R}^H$ is a transformation defined by a rotation matrix $R \in \mathbb{R}^{H \times H}$, a translation vector $t \in$

$\mathbb{R}^H$ and uniform scaling $s \in \mathbb{R}$: $f_{t,R,s}(x) = sRx + t$. The inverse of a similitude is also a similitude: $f_{s,R,t}{}^{-1}(x) = f_{\frac{1}{s},R^\top,-\frac{1}{s}R^\top t}(x)$.

$f_{s,R,t}(\mathcal{N}_0)$ is a spherical MVN with mean t and covariance s^2I. Transforming a generic spherical MVN by a similitude can be cast as the transformation of $\mathcal{N}_0$ by two similitudes:

$$f_{t,R,s}(\mathcal{N}_{t_0,s_0^2 I})(x) = f_{t,R,s}(f_{t_0,R_0,s_0}(\mathcal{N}_0))(x)$$

$$= (2\pi)^{-\frac{H}{2}}(ss_0)^{-H}\exp\left[-\frac{1}{2s_0^2 s^2}\|x - (sRt_0 + t)\|^2\right] \quad (6.1)$$

The Expectation-Maximization (EM) Algorithm   [35]

Let $p(X \mid Z, \theta)$ be a probability distribution with parameters θ and latent variables Z. In the most common example of the EM algorithm, p is a mixture of MVNs: θ contains the parameters of K MVNs together with a weight w_k for each. Z is a binary matrix whose rows z_i determine which component is responsible for x_i: i.e. if component k is responsible, then Z_{ik} is 1 and all other elements of row i are 0. The maximum likelihood parameters for a dataset X are determined by

$$\arg\max_\theta \ln p(X|\theta) = \arg\max_\theta \ln \sum_Z p(X|Z,\theta)p(Z)$$

where the sum is over all possible values of Z. The value inside the sum is the *complete-data likelihood* $p(X, Z \mid \theta)$. In the mixture-of-MVNs example, there are K^N possibilities for Z, making this sum intractable for practical data.

The intuition behind the EM algorithm is to optimize θ with respect to our best guess for Z and vice versa. Let θ^{old} be our current best guess for the parameters. We first optimize $p(Z \mid X, \theta^{old})$, the posterior on the latent variables. Note that, in the MVN example, we can represent this as an $N \times K$ matrix Z with $Z_{ij} = p(j \mid x, \theta) = N_j^{old}(x)w_j$. These are known as the *responsibilities*.

We then determine the expectation of the logarithmic complete-data likelihood under our posterior $p(Z|X, \theta^{old})$ as a function of θ.

$$Q(\theta) = \sum_Z p(Z|X, \theta^{old}) p(Z) \ln p(X, Z \mid \theta) \qquad (6.2)$$

Note that the complete-data likelihood is now inside the logarithm. We optimize this Q-*function*, with respect to θ:

$$\theta^{new} = \arg \max_\theta Q(\theta).$$

The computation of the matrix Z is know as the *expectation* step and the computation of θ^{new} is known as the *maximization* step. The EM algorithm iterates the two steps until the parameters converge.

6.2   The IFS model

We will define an iterated function system as follows:

Definition 6.1. An *Iterated Function System* of order K and dimension H is a pair $(\{f_k\}, \{w_k\})$ of K *components* $f_k : \mathbb{R}^H \to \mathbb{R}^H$ with K associated *weights* w_k, nonnegative scalars, with $\sum_i w_i = 1$.

Here, all components are similitudes: i.e. f_k is defined by a scalar $s_k \in \mathbb{R}$, a rotation matrix $R_k \in \mathbb{R}^{H \times H}$ and a translation vector $t_k \in \mathbb{R}^H$:

$$f_k(x) = s_k R_k x + t_k.$$

The definition can be extended to functions on any kind on metric spaces [51]. As noted in the introduction, similitudes provide a good balance between expressive power, and ease of optimization.

An IFS determines a probability distribution on $\mathbb{R}^H$. We can sample from this distribution as follows. Let p_0 be some initial distribution on $\mathbb{R}^H$, whose support is a compact set, and let D be some nonnegative integer. Sample x_0 from p_0. Sample a random component, so that component f_k has probability w_k. Let $x_2 = f_k(x_1)$. Continue this process until x_D. x_D was sampled from the depth-D distribution defined by the IFS.

We note two important properties of IFSs. First, the distribu-

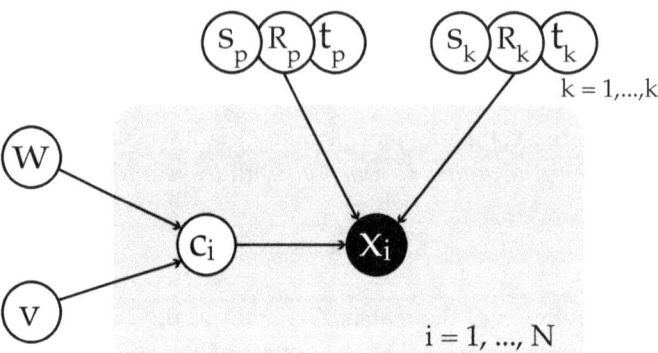

Figure 6.2: A graphical model, illustrating the components of the IFS model. The gray box is a plate, representing a repetition of the nodes inside the box for each datapoint. The black node represents the observed data.

tion on x_D converges with increasing D: that is, the KL divergence between successive models goes to zero with D. We call the distribution it converges to the *limit distribution*. Second, the limit distribution is independent of the choice of p_0, so long as p_0 has compact support. Thus, we can take the weights and components as *parameters* that determine a distribution, and we can evaluate the IFS to high D to approximate this distribution. For formal statements of these properties and their proofs we refer the reader to [51].

To make the model easier to fit to natural data, which may not be a perfect fractal, and may not be properly centered, we extend the basic IFS model with some additional factors. This complete model is illustrated in Figure 6.2. It consists of:

the components $f_k = \{s_k, \mathbf{R}_k, \mathbf{t}_k\}$ K similitudes, each determined by $s_k \in \mathbb{R}$, $\mathbf{t}_k \in \mathbb{R}^H$, $\mathbf{R}_k \in \mathbb{R}^{H \times H}$.

the component weights w This is a vector of K non-negative values, summing to 1. Each value determines the mixture-weight of its associated component in the iteration of the IFS. That is, if p_0 is the density function of the initial distribution, the model after one iteration is the mixture $p_1(x) = \sum_k w_k f_k(p_0)(x)$.

the depths v Instead of assuming that all points come from the limit distribution of the IFS, we assume that the depth to which the model is evaluated is variable, and differs per point. Let D be the maximum depth (a parameter of the algorithm). v is a length-D vector on nonnegative values, summing to 1, with v_d the probability that a point is generated by the model at depth d.

The post-transformation s_p, R_p, t_p The parameters for a single *post-transformation*. This transformation is applied once to all sampled points.

The code c_i An ordered sequence $c_i = \langle c_{i1}, \ldots, c_{id} \rangle$. Each element of the code is an integer in $[1, K]$ representing a component.

The data x_i This is the point after the post-transformation is applied: i.e. the observed data.

The first four items, those outside the plate in Figure 6.2, form the *parameters* of the model. We will refer to these combined parameters of a single model as θ. Inside the plate are the *observed* variables (the data) and the *latent* variables (the codes).

For the initial model, we choose the standard multivariate Normal distribution $\mathcal{N}_0$, mentioned in the preliminaries. Since each affine transformation of an MVN is itself an MVN, this makes each iteration of our model a mixture of MVNs. Since the depth vector v mixes these models, the whole model is also a mixture of MVNs. Each MVN in this mixture is determined by a sequence of components with length between 0 and D; its *code*. We will use the notation $[1, K]^d$ to refer to the set of all length-d codes. Let $[1, K]^{[a,b]} = \bigcup_{d \in [a,b]} [1, K]^d$. Thus, the set of all codes $\{c_1, \ldots, c_M\}$, including the empty code, is $[1, K]^{[0,D]}$. Given a code c_j, we define the function $f_j : \mathbb{R}^H \to \mathbb{R}^H$ as the composition of the post-transform and the components indicated by the code:

$$f_j = f_p \circ f_{c_{j1}} \circ \ldots \circ f_{c_{jd}}.$$

Using this notation, we can write the probability density function of the model as follows:

$$p(x; \theta) = \sum_{j=1}^{|[1,K]^{[0,D]}|} p(c_j; \theta) f_j(p_0)(x)$$

$$= \sum_j v_{|c_j|} \prod_i w_{c_{ji}} f_j(\mathcal{N}_0)(x).$$

Giving the model a variable depth has several advantages. First, it makes the model a generalization of good fall-back models: with $v_0 = 1$, the model becomes a spherical MVN, determined by the post-transform. With $v_1 = 1$, the model becomes a mixture of K spherical MVNs.[2] Since each point has its own depth, the full model is a mixture over these two models, and the deeper IFS models. If the data is not self-similar, or only partially self-similar, models with high weights on the lower depths can account for this.

The variable depth also allows the EM search a "gentle start". In the limit, most IFSs have a support with lower dimension than the embedding dimension of the data: i.e. almost all of $\mathbb{R}^H$ has probability zero. This means that even if the model fits the source of the data perfectly, the slightest addition of noise will make the entire likelihood zero. It also means that a minute change in parameters can mean the difference between the maximum likelihood, and likelihood zero. In effect, for high values of d, the fitness landscape becomes very jagged.

The depth parameters provide an automatic defense: if such a drop occurs, the lower models, whose fitness landscape is smoother, automatically get a higher weight, smoothing out the fitness landscape for the complete mixture. As the search converges to the correct model, the model becomes more weighted towards the higher depths.

The variable depth also creates some complications, specifically when our data is not perfectly centered. If we transform an IFS, evaluated to depth d, by an affine transformation, the result is also an IFS. However, the resulting components depend on d. Imagine a dataset sampled from a Sierpinski triangle translated away form the origin. We have good models at every depth for

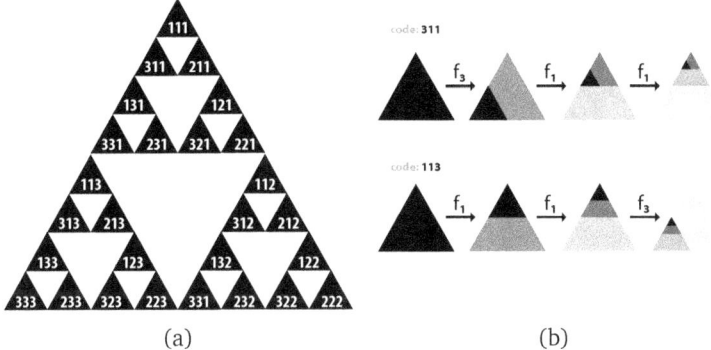

(a)   (b)

Figure 6.3: (a) Codes of length three on the Sierpinski triangle and the subsets they code for. (b) The construction of a subset from its code.

this data, but their components are different. Only if the data is centered do the components coincide for all depths. This is desirable, because once the solution has converged to these components, we will see the higher depths gaining more weight. But how the data should be centered for the components to coincide differs from one IFS to another. This is the reason for the *post-transform*: we learn a centered IFS, and transform it to the data.

6.3   The EM algorithm for IFS models

The basic idea of the algorithm is to assign *codes* to points in the data. To illustrate this, consider the Sierpinski triangle, as shown in Figure 6.3. Each application of a distinct, finite sequence of components maps the initial triangle to a disjoint region. Whether these regions are disjoint depends on the initial image and the IFS, but since it helps to illustrate the point, we have chosen those here so that they are. We can label each region with the specific sequence of components that mapped the initial triangle to that point: its code. Note that finding a code corresponding to a particular endpoint can be counter-intuitive, as the first component applied determines the location at the smallest scale, and the last component applied determines the location at the largest scale.

---

[2]To generalize to non-spherical MVNs and mixtures over non-spherical MVNs, the component family should be extended to positive definite affine functions.

Given a model like the Sierpinski triangle, it's a straightforward matter to find the best code for a given point. Conversely,
given the best code for each point, we can reconstruct the transformations, since we know which regions of space map to one
another under each transformation. This is the basic idea of the
algorithm: we iterate these two steps. We find the codes given
some initial model, reconstruct a new model from the codes and
we iterate this process.

Finding the codes given a model is called the *expectation* step,
and finding the model given the codes is called the *maximization*
step. We will detail both steps below.

6.3.1 Expectation: the latent variables

Instead of assigning a single code to a single point, however, we
make this step probabilistic: each code takes a certain amount
of *responsibility* for every point x. The responsibility taken by
component j for point i is $p(c_j \mid x_i) \propto p(x_i \mid c_j)p(c_j)$. These
factors correspond to:

$$p(x \mid c_j) = \mathcal{N}(x; \mu_j, \Sigma_j)$$

$$p(c_j) = p(d) \prod_i p(c_j) = v_{|c_j|} \prod_{i \in c_j} w_i$$

Where μ_j is the mean corresponding to $f_j(\mathcal{N}_0)$ with $f_j = f_p \circ$
$f_{c_{j1}} \circ f_{c_{j2}} \circ \ldots \circ f_{c_{ja}}$, and Σ_j is the covariance matrix $f_j(\mathcal{N}_0)$.

Let $M = |[1, K]^{[0,D]}|$ and let P be an $N \times M$ matrix with:

$$P_{ij} = p(c_j \mid x_i) = \frac{p(x_i \mid c_j)p(c_j)}{\sum_{a \in [1,K]^{[0,D]}} p(x_i \mid c_a)p(c_a)}.$$

The size of this matrix will grow very fast with D, and most
of its values will be practically indistinguishable from 0 when
normalized. For this reason, it may be advisable to set all but the
largest values to 0, and store the matrix in a sparse datastructure.
For our experiments, such optimizations were not necessary.

6.3.2 Maximization: the parameters

This section provides only the equations for each parameter, with a brief motivation. See the appendix for more detailed derivations.

Let θ^{old} be the model we used to compute the responsibilities. Filling in the probability functions for our data and latent variables, the Q-function (6.2) becomes:

$$Q(\theta) = \sum_i p(z_i \mid x_i, \theta^{\text{old}}) \ln p(x_i \mid z_i, \theta) p(z_i \mid \theta)$$

$$= \sum_{i=1,j=1}^{N,M} P_{ij} \ln \mathcal{N}_j(x) \, v_{|c_j|} \prod_{a \in c_j} w_a \,.$$

The maximization step consists in optimizing Q for the different parts of θ. In most cases, we can achieve this by taking the partial derivative and setting it equal to zero. For clarity of notation, when optimizing for a certain subset q of the parameters θ (such as the depths, weights or parts of the components) we will write $Q(q)$, and omit any terms that are constant with respect to q. We may also omit any constant multiplier of the whole function.

The depths v_d To find the optimal depth weights we solve $\partial Q(v)/\partial v = 0$, using a Lagrange multiplier to enforce that v sums to one, which gives us:

$$\hat{v}_d = \frac{p^d}{\sum_i p^i} \text{ with } p^d = \sum_{i,j:|c_j|=d} P_{ij}$$

The components f_k **and weights** w_k Even if we optimize the latent variables and the parameters separately, the optimization of f_k and f_p is difficult. We must find K maps and weights, and a post-transformation f_p, such that all the M endpoint distributions provide maximal likelihood to their assigned points. The problem is that each term in Q is a complicated mix of multiple components.

To make the optimization of Q practical, we simplify the task in two ways. First, we optimize f_p and $(\{f_k\}, \{w_k\})$ separately, taking the parameters not being optimized from θ^{old}. Second, we simplify the Q function. We first rewrite it as follows: Let kc_j be the code $\langle k, c_{j1}, \ldots \rangle$ and let $M^- = |[1, K]^{[0, D-1]}|$. Let $Y = f_p^{old-1}(X)$. Then:

$$Q(\{f_k\}, \{w_k\})$$

$$= \sum_k \sum_{i=1, j=1}^{N, M^-} p(kc_j \mid x_i, \theta^{old}) \ln f_p f_k(\mathcal{N}_j)(x_i) p(c_j \mid \theta^{old}) p(k)$$

$$= \frac{1}{s_p} \sum_k \sum_{i,j} P_{ij}^k \ln \left[f_k(\mathcal{N}_j)(y_i) \, v_{|c_j|+1} p(c_j) \, w_k \right].$$

Where we have defined a submatrix P^k of P, which contains all columns corresponding to codes beginning with the symbol k:

$$P_{ij}^k = p(kc_j \mid x_i) = \frac{p(x_i \mid kc_j) p(kc_j)}{\sum_{a \in [1, K]^{[0, D]}} p(x_i \mid c_a) p(c_a)}.$$

We have now written Q in a form that isolates only the first component in the code. Of course, each $\mathcal{N}_j$ in this sum still depends on all the components in the code c_j, but this is where we simplify the function: we take $\mathcal{N}_j$ to be a *constant*, computed from θ^{old}, and optimize only for the first component in the code. This gives us, for the k-th component:

$$Q(f_k, w_k) = \sum_{i,j} P_{ij}^k \ln f_k(\mathcal{N}_j)(y_i) + \sum_{i,j} P_{ij}^k \ln w_k$$

Finding the optimal w_k follows the same principle as the depths: we take the derivative, set it equal to 0, and use a Lagrange multiplier for the constraint. We find:

$$\hat{w}_k = p^k / \sum_i p^i.$$

with $p^k = \mathbf{1}^T P^k \mathbf{1}$, the sum of the elements of P^k.

If we isolate f_k, the problem is very similar to the one solved to construct the Coherent Point Drift (CPD) algorithm [71]: transform a given set of MVNs to maximize the likelihood of a dataset, with respect to responsibilities $\mathbf{P}$. The main difference is that in our situation, each component N_j has its own variance. We follow the same approach, and as we shall see, the solutions for our problem are very similar to those of the CPD algorithm.

Using Equation 6.1 from the preliminaries, we can rewrite $Q(f_k)$ as a mixture of transformations of $\mathcal{N}_0$:

$$Q(s_k, \mathbf{R}_k, \mathbf{t}_k) = -p^k H \ln s_k - \sum_{i,j} P^k_{ij} \frac{1}{2s_j{}^2 s_k{}^2} \|y_i - \mathbf{t}_k - s_k \mathbf{R}_k \mathbf{t}_j\|^2$$

where s_j, $\mathbf{R}_j$ and $\mathbf{t}_j$ are the parameters of the similitude $f_p \circ c_{j1} \circ \ldots \circ c_{jd}$.

To find the optimal translation $\mathbf{t}_k$, we solve $\partial Q(\mathbf{t}_k)/\partial \mathbf{t}_k = 0$, which yields, in matrix notation:

$$\hat{\mathbf{t}}_k = \frac{1}{p^k_z} \mathbf{Y}\mathbf{P}^k \mathbf{Z}\mathbf{1} - \frac{1}{p^k_z} s_k \mathbf{R}_k \mathbf{T}\mathbf{Z}\mathbf{P}^{k\,\mathsf{T}}\mathbf{1} = \mathbf{y}^k - s_k \mathbf{R}_k \mathbf{t}^k$$

where $\mathbf{T}$ is the matrix with $\mathbf{t}_j$ as its columns,

$$\mathbf{Z} = \mathrm{diag}(s_1{}^{-2}, \ldots, s_M{}^{-2})$$

and $p^k_z = \mathbf{1}^\mathsf{T}\mathbf{P}^k\mathbf{Z}\mathbf{1}$. Thus $\hat{\mathbf{t}}_k$ is the difference between a weighted mean of the data $\mathbf{y}^k$ and a weighted mean of the endpoint means $\mathbf{t}_j$, scaled and rotated, where in both cases, the matrix $\mathbf{P}^k\mathbf{Z}$, normalized to sum to one, determines the weights. Note that this is not a complete solution, since it still depends on s_k and $\mathbf{R}_k$. We can, however, plug $\hat{\mathbf{t}}_k$ back into the Q-function to optimize for $\mathbf{R}_k$.

Finding the optimal rotation matrix $\mathbf{R}_k$ is more complex than simply finding the derivative and setting it equal to zero, since we have the constraint that $\mathbf{R}_k$ is orthogonal and has determinant 1. We use the technique described in [70] and [71, Lemma 1]. We first rewrite the objective function to the form $\mathrm{tr}(\mathbf{A}^\mathsf{T}\mathbf{R}_k)$, for some

A. We fill in $\hat{t}_k$, and reduce to

$$Q(\mathbf{R}_k) = \mathrm{tr}\left(\left[\mathbf{Y}^k\mathbf{P}^k\mathbf{Z}\mathbf{T}^{k^\mathsf{T}}\right]^\mathsf{T}\mathbf{R}_k\right) \quad \text{with} \quad \begin{cases} \mathbf{Y}^k &= \mathbf{Y} - \mathbf{y}^k\mathbf{1}^\mathsf{T} \\ \mathbf{T}^k &= \mathbf{T} - \mathbf{t}^k\mathbf{1}^\mathsf{T} \end{cases}$$

The optimal $\mathbf{R}_k$ can be derived from the singular value decomposition (SVD) of $A = \mathbf{Y}^k\mathbf{P}^k\mathbf{Z}\mathbf{T}^{k^\mathsf{T}}$: if $A = \mathbf{U}\mathbf{S}\mathbf{V}^\mathsf{T}$, with $\mathbf{U}$, $\mathbf{S}$ and $\mathbf{V}$ defined as normal for the SVD then we have $\hat{\mathbf{R}}_k = \mathbf{U}\,\mathrm{diag}(1,\ldots,1,|\mathbf{U}\mathbf{V}^\mathsf{T}|)\,\mathbf{V}^\mathsf{T}$.

Finally, we derive the scaling parameter s_k by filling in $\hat{t}_k$ and solving $\partial Q(s_k)/\partial s_k = 0$. We get:

$$0 = s_k^{-2}\,\mathrm{tr}(\mathbf{Y}^{k^\mathsf{T}}\mathrm{diag}(\mathbf{P}^k\mathbf{Z}\mathbf{1})\mathbf{Y}^k) + s_k^{-1}\,\mathrm{tr}(\mathbf{T}^k\mathbf{Z}\mathbf{P}^{k^\mathsf{T}}\mathbf{Y}^{k^\mathsf{T}}\mathbf{R}_k) - \mathrm{H}\mathrm{p}^k\,.$$

This is a quadratic equation in s_k^{-1}, which we can solve and invert to find $\hat{s}_k$.

The post transform: s_p, $\mathbf{R}_p$, t_p We now fix the components f_k, considering them constant and taking their values from θ^{old}, and optimize for the parameters of f_p. The Q-function becomes:

$$Q(s_p, \mathbf{R}_p, t_p) = \sum_{i=1,j=1}^{N,M} p(c_j \mid x_i)\ln f_p(N_j)(x_i)p(c_j)$$

$$= -p\ln s_p - \frac{1}{2s_p{}^2}\sum_{i,j} s_j^{-2}P_{ij}\|x - t_p - s_p\mathbf{R}_p t_j\|^2$$

where s_j and t_j are constants derived from θ^{old}. Note that j now iterates over all codes.

The form of this Q function is the same as the ones we used to derive the optimal components f_k. We follow the same derivations and get:

$$\hat{t}_p = x - s_p\mathbf{R}_p t^p$$

$$\hat{\mathbf{R}}_p = \mathbf{U}\,\mathrm{diag}(1,\ldots,1,|\mathbf{U}\mathbf{V}^\mathsf{T}|)\,\mathbf{V}^\mathsf{T}$$

$$\text{with } \mathbf{U}\mathbf{S}\mathbf{V} = \mathrm{svd}(A), \ A = \mathbf{X}^p\mathbf{P}\mathbf{Z}\mathbf{T}^{p^\mathsf{T}}$$

with

$$x^p = (\mathbf{1}^T \mathbf{PZ1})^{-1} \mathbf{XPZ1} \qquad\qquad t^p = (\mathbf{1}^T \mathbf{PZ1})^{-1} \mathbf{TZP}^T \mathbf{1}$$

$$\mathbf{X}^p = \mathbf{X}^p - x^p \mathbf{1}^T \qquad\qquad \mathbf{T}^p = \mathbf{T} - t^p \mathbf{1}^T$$

$$\mathbf{Z} = \mathrm{diag}(s_1^{-2}, \ldots, s_M^{-2})$$

Finally, for s_p, we solve

$$0 = s_p^{-2} \mathrm{tr}(\mathbf{X}^T \mathrm{d}(\mathbf{PZ1})\mathbf{X}) + s_p^{-1} \mathrm{tr}(\mathbf{TZP}^T \mathbf{X}^T \mathbf{R}_p) - \mathrm{Hp}$$

6.3.3 Dealing with singularities

In the EM algorithm for MVN mixtures, there is a danger that the algorithm becomes stuck in a situation where one or more of the components do not have responsibility for any part of the data, or for only a single point. In this case, the covariance matrix for such components cannot be computed, and the algorithm must be reset in some way.

In the IFS algorithm a similar thing can happen, causing one or more of the matrices $\mathbf{P}^k$ to become low-rank. In this case, the SVD decomposition required to find $\mathbf{R}_k$ cannot be computed. When this happens, we use the following strategy: we remove the undetermined component, and for each one we *split* one of the well-determined components. Let $f_a = (s_a, \mathbf{R}_a, t_a)$ with weight be the well-determined component and f_b be the singleton component. We resolve the situation by setting:

$$f_a \leftarrow (s_a, \mathbf{R}_a, t_a + \epsilon_a)$$

$$f_b \leftarrow (s_a, \mathbf{R}_a, t_a + \epsilon_b)$$

where ϵ_a, ϵ_b are vectors with elements randomly drawn from $\mathcal{N}(0, \sigma)$, where we use $\sigma = 0.01$ in all experiments. The old weight w_a is distributed equally over the components f_a and f_b and the weight vector is re-normalized.

The rationale is easiest to understand if we take $v_1 = 1$ and view the model as a mixture of Guassians. In that case, the well-determined components cover all the data. By splitting f_a, and adding some small noise, the points formerly 'claimed' by f_a will now be distributed approximately evenly between f_a and f_b.

A second trap is that the variance of the endpoint distributions can become so small that, even using logarithmic representation, all matrices P^k become low-rank (ie, all entries are 0 for certain components). If this happens, we reset the algorithm by approximating P and all P^k: we still use the endpoint means t_j from the model, but we place an MVN with a fixed standard deviation on each (0.01 in all experiments) and re-compute the responsibilities under that model.

6.4   Results

To speed up the algorithm, we use a subsample (with replacement) from the data in all experiments. After each iteration (one expectation and one maximization step), we draw another sample. The sample size is always 500. We run the algorithm for 300 iterations, with maximum depth 6. In practice, it may be advisable to increase the sample size and the depth as the algorithm begins to converge, but for the sake of simplicity, we have kept the parameters fixed.

For our initial model we use the following process: we choose K points $\{z_k\}$ uniformly at random on the bi-unit sphere. We then construct for each of these points, a similitude f_k with $s_k = 0.5$, and a uniform random rotation, choosing t_k so that f_k's fixed point is z_k. This results (for K = 3) in a kind of Sierpinski triangle covering a decent amount of the space occupied by the data. v and w are initialized with all values equal.

6.4.1 Synthetic distributions

Figure 6.4 shows the performance of the algorithm on four well-known fractals: the Sierpinski gasket, a Sierpinski gasket with unequal weights, and two- and four-component versions of the Koch curve.

For each dataset we repeated the experiment 100 times and show both the model with its learned mixture over depths, and what the IFS looks like when evaluated to infinite depth.[3] We

data	it. 20	it. 40	final model	full depth

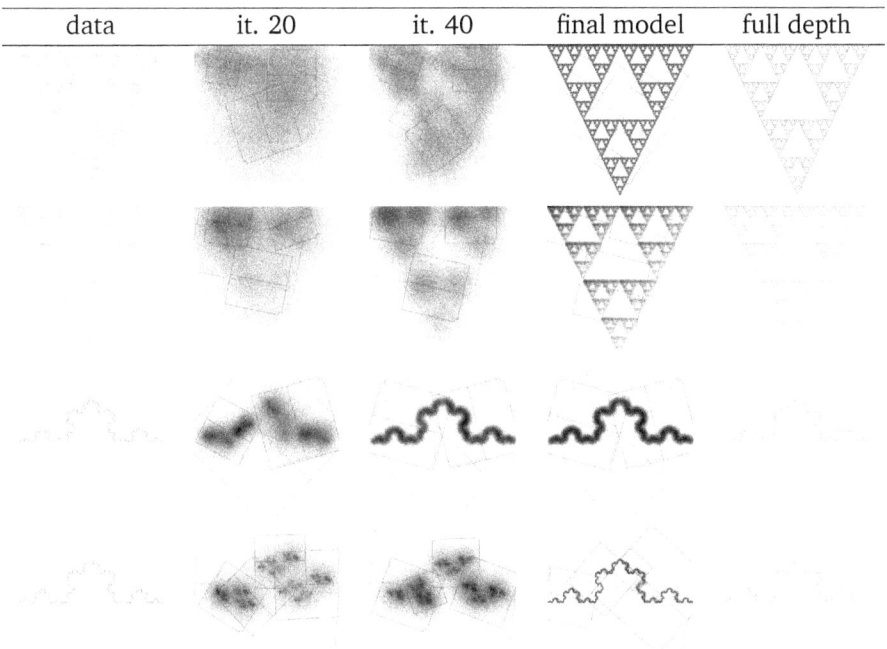

Figure 6.4: Results of EM search for known models. The boxes show the model: the post-transform maps the image frame onto the larger box, and each component maps the larger box onto one of the smaller boxes. The bars in the side of the blue boxes show the weight of each component. The learning tasks, from top to bottom are: the Sierpinski gasket, the Sierpinski gasket with unequal weights, the Koch curve with 4 components and the Koch curve with two components.

show the run that ended with a model with the greatest likelihood (on the training data). the end results of all runs are shown in the appendix.

It is difficult to objectively quantify the number of trials that successfully converged to the required IFS. One may suggest testing the likelihood of the data under the learned model, to see if it is close to that of the target model, but the likelihood grows exponentially as the higher depths get greater weight (if the model is correct). This means that unless the algorithm finds the absolute correct model, the likelihoods of any learned model, correct or otherwise, will be much closer to one another than to the likelihood of the target model.

We inspected the resulting models for each of the 100 runs visually to determine whether they were good approximations, and report the proportions here, with the proviso that these necessarily include some level of subjective judgment. The complete set of results for each experiment is reproduced in the appendix. We estimate that the following proportion of experiments ended in the neighbourhood of one of the global optima:

Sierpinski 67%

Sierpinski, unequal weights 51%

Two-component Koch 20%

Four-component Koch 23%

In all cases, we can see in the complete results that there are many different ways to achieve the same limit set, especially with the extra degrees of freedom introduced by the addition of the post transform. We also observe that if the component weights in the source of the data are uneven, this makes the learning task more difficult. The more uneven the weights become, the more likely the model is to get stuck in a local optimum. We have also found that working with centered data, and eliminating the post transform drastically improves results, but how the data should be centered is only known if the source of the data is known. Therefore, such experiments are not representative of a realistic parameter estimation scenario.

---

[3] Such data can be sampled using an algorithm known as the *chaos game* [17].

6.4.2 Non IFS data

Next, we try the model on five non-IFS datasets. The first three are derived from images of fractal phenomena: a Romanesco broccoli, a coast line, and the edge of a cloud, illuminated by sunshine. While these are fractals, they also contain a degree of noise, and a deterministic IFS is unlikely to capture them perfectly. The fourth dataset is sampled from a uniform distribution on the bi-unit disc. Unlike the square, the disc cannot be described as an IFS, so the deeper models will likely leave "gaps" of low probability density. At low depths, however, the model is forced to spend probability mass on the area outside the disc, so some balance must be struck.

While none of these datasets are IFSs, the first three do contain self-similarity, and we see that the models are using the higher depths of the IFS to capture that. In the case of the circle, we see that the learned model looks very different from a circle (which cannot be described as the limit set of an IFS), but there is nevertheless an IFS that places probability mass in the correct regions.

6.4.3 Complexity

To give an indication of the complexity of the algorithm, we plot the time taken for a single expectation step and a single maximization step, against three variables: the size of the data sample, the dimension of the data, and the maximum depth.

In the first case, we sample the data from a three-dimensional standard normal distribution, use the Sierpinski gasket as an initial model, and compute one full iteration of the algorithm, measuring the time taken to complete the expectation and the maximization step. The depth is set to 5.

For the second test, we again sample from a standard normal distribution, but we vary its dimension from 1 to 50. The model is initialized at random using the method described earlier. The depth is set to 5.

For the final test we use the same procedure as for the first, but we fix the data size at 250 and vary the depth from 0 to 10.

For all values, the experiments were repeated 100 times. The

data	it. 20	it. 40	final model	full depth

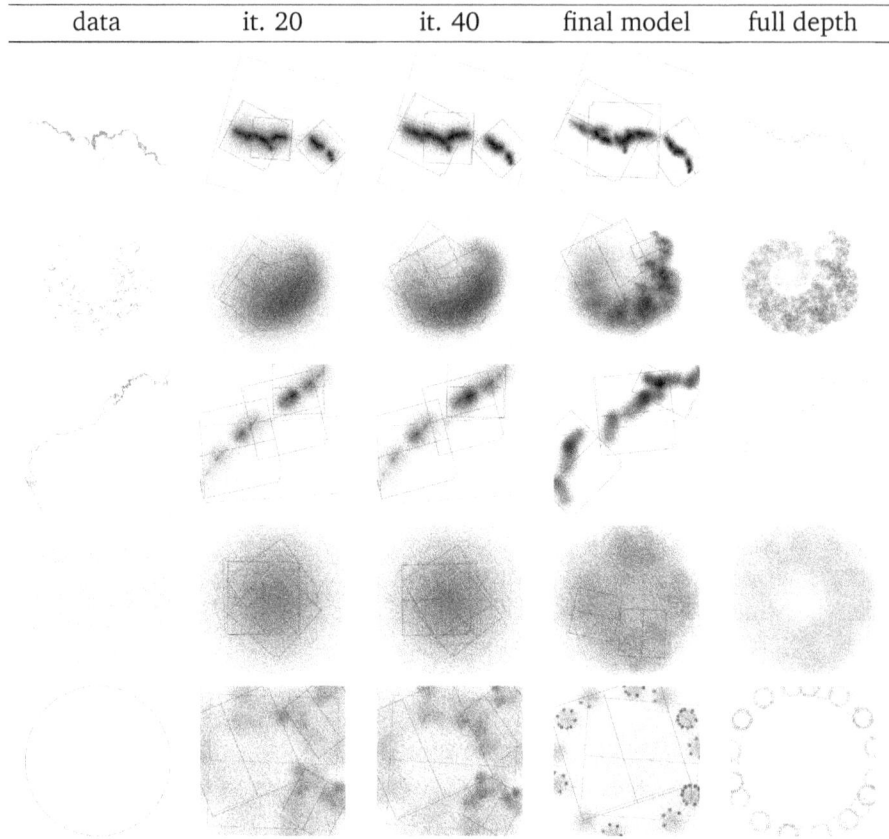

Figure 6.5: Results of EM search for known models. The red and blue boxes show the model: the post-transform maps the image frame onto the red box, and each component maps the red box onto a blue box. The bars in the side of the blue boxes show the weight of each component. The learning tasks, from top to bottom are: the outline of a cloud, shadows cast on a romanesco broccoli, a small part of the Australian coast, a uniform distribution on the bi-unit disc and a uniform distribution on a circle.

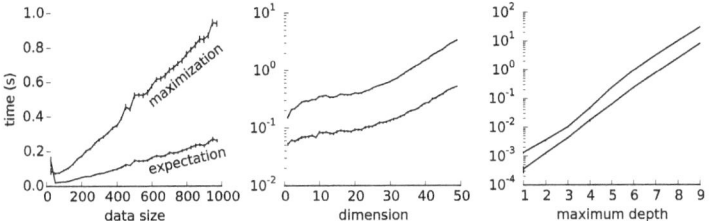

Figure 6.6: The time taken for a single iteration of the algorithm, separated into the expectation and the maximization step. For each value, the iteration was computed one hundred times. The graph shows the mean over these repetitions, the error bars show a 95% confidence interval (i.e. 1.96 times the standard error on both sides). Note the logarithmic axes on the second and third plot.

order of the experiments was randomized over all values, repeats and variables, to eliminate any temporary effects, like increased network load. The experiments were performed on a single machine with 8 Gigabyte of java heapspace and 2 1.80 Ghz Intel Xeon processors (E5-2650L). The code for a single iteration is single-threaded, and the matrix operations were not hardware accelerated.[4] These options are still open to improve the algorithm's efficiency. The fast Gauss transform [45, 95] may also speed up the computation of the expectation step.

Figure 6.6 shows the result. The strongest growth is in the depth, which is at least exponential, as the number of columns of P alone grows exponentially in D. The growth in dimension looks to be polynomial first, up to around 20, and exponential afterwards. Why this is the case, and whether a polynomial growth for all dimensions is achievable requires further analysis.

6.5 Discussion

We have introduced a new algorithm for the induction of fractal models. To our knowledge, this is the first such algorithm that does not use a general-purpose optimization technique like genetic algorithms.

---

[4] Specifically, we used the Apache Commons Mathematics library.

Similitudes were chosen as a nice balance between expressiveness and parameter complexity. More general function classes are certainly possible, although deriving the maximization step analytically may not be possible for these, and convex optimization or even locally optimal solutions may be required in the maximization step.

An alternative approach would be to use a variational Bayesian algorithm [19] instead of an EM algorithm. If a variational optimization algorithm exists for the desired function class, it should be straightforward to plug this into a general variational algorithm for iterated function systems. Such an approach would also avoid the problem of singularities, and it would allow some tuning of the algorithm through the selection of priors. We consider this a promising direction for future research.

Many natural fractal phenomena are not precisely captured by iterated functions systems. Coastlines, trees and clouds are self-similar, but in a much more random manner than IFSs can describe. One solution takes the form of an extension to random iterated function systems, as described in [48]. Here, at each application the IFS is itself chosen at random from a distribution on IFSs. Another option would be to manually add domain knowledge about the data into the model, for instance, specific knowledge about the development of clouds or the growth of trees. In both cases, we believe the EM algorithm provides a basic template for a solution: the main issue is that of finding the specific choices made inside the model to arrive at each point in the dataset. By casting this sequence of choices as a latent variable, we can divide and conquer: we cannot solve the problem as a whole, but we can find the latent variables given the parameters, and the parameters given the latent variables. Our original motivation was the case of self-similar graphs. The graph-analogue of an IFS is a *Kronecker graph*: Kronecker graphs have been seriously studied as models for real-world graph data [60]. Currently, these models are fit by a gradient descent algorithm. The latent-variable approach described here, may translate to algorithms for fitting Kronecker graphs. As noted in the introduction, most research on large graphs implicitly assumes some level of self-similarity in the graph under study. Models like these make this assumption explicit, and may lead to new insight into the structure of large graphs.

Algorithm 2 One iteration of the IFS-EM algorithm.

Given: a dataset $\mathbf{X}$, a number of components K, a maximum depth D.

$\mathbf{P}_{ij} \leftarrow \mathcal{N}_j(\mathbf{x}_i)\mathbf{v}_{|c_j|} \prod_{a \in c_j} \mathbf{w}_a$ *# Expectation step*

Normalize $\mathbf{P}$ so that $\mathbf{P1} = \mathbf{1}$

for each $k \in [1, K]$:

    Let $\mathbf{P}^k$ be the submatrix of $\mathbf{P}$'s columns j for which $c_{j1} = k$

$\mathbf{Y} \leftarrow f_p^{\text{old}-1}(\mathbf{X})$ *# Maximization step*

for all d, $\hat{\mathbf{v}}_d \propto \sum_{j:|c_j|=d} \mathbf{P}_{ij}$

for each $k \in [1, K]$:

    $\hat{w}_k \leftarrow \mathbf{1}^\mathsf{T}\mathbf{P}^k\mathbf{1}/\sum_i \mathbf{1}^\mathsf{T}\mathbf{P}^i\mathbf{1}$

    $\mathbf{y}^k \leftarrow p^{k-1}\mathbf{Y}\mathbf{P}^k\mathbf{Z1}, \quad \mathbf{t}^k \leftarrow p^{k-1}\mathbf{TZP}^{k\mathsf{T}}\mathbf{1}$

    $\mathbf{Y}^k \leftarrow \mathbf{Y} + \mathbf{y}^k\mathbf{1}^\mathsf{T}, \quad \mathbf{T}^k \leftarrow \mathbf{T}^k + \mathbf{t}^k\mathbf{1}^\mathsf{T}$

    $\mathbf{Z} \leftarrow \text{diag}(s_1^{-2}, \ldots, s_M^{-2})$

    $\mathbf{U}, \mathbf{S}, \mathbf{V}^\mathsf{T} \leftarrow \text{svd}(\mathbf{Y}^k\mathbf{P}^k\mathbf{Z}\mathbf{T}^{k\mathsf{T}})$

    $\hat{\mathbf{R}}_k \leftarrow \mathbf{U}\,\text{diag}(1, \ldots, 1, |\mathbf{U}\mathbf{V}^\mathsf{T}|)\mathbf{V}^\mathsf{T}$

    $\hat{s}_k$: solve $s_k^{-2}\,\text{tr}(\mathbf{Y}^{k\mathsf{T}}d(\mathbf{P}^k\mathbf{Z1})\mathbf{Y}^k) + s_k^{-1}\,\text{tr}(\mathbf{T}^k\mathbf{Z}\mathbf{P}^{k\mathsf{T}}\mathbf{Y}^{k\mathsf{T}}\mathbf{R}_k) - Hp^k = 0$

    $\hat{\mathbf{t}}_k \leftarrow \mathbf{y}^k - s_k\mathbf{R}_k\mathbf{t}^k$

$\mathbf{X}^p \leftarrow \mathbf{X} - \mathbf{x}^p\mathbf{1}^\mathsf{T}, \quad \mathbf{T}^p \leftarrow \mathbf{T} + \mathbf{t}^p\mathbf{1}^\mathsf{T}$

$\mathbf{Z} \leftarrow \text{diag}(s_1^{-2}, \ldots, s_M^{-2})$

$\mathbf{U}, \mathbf{S}, \mathbf{V}^\mathsf{T} \leftarrow \text{svd}(\mathbf{X}^p\mathbf{P}\mathbf{Z}\mathbf{T}^{p\mathsf{T}})$

$\hat{\mathbf{R}}_p \leftarrow \mathbf{U}\,\text{diag}(1, \ldots, 1, |\mathbf{U}\mathbf{V}^\mathsf{T}|)\mathbf{V}^\mathsf{T}$

$\hat{s}_p$: solve $s_p^{-2}\text{tr}(\mathbf{X}^\mathsf{T}d(\mathbf{PZ1})\mathbf{X}) + s_p^{-1}\text{tr}(\mathbf{TZP}^\mathsf{T}\mathbf{X}^\mathsf{T}\mathbf{R}_p) - Hp = 0$

$\hat{\mathbf{t}}_p \leftarrow \mathbf{x}^p - s_p\mathbf{R}_p\mathbf{t}^p$

THE SINGLE SAMPLE SETTING

We began our introduction with an esoteric scenario: a scientist faced with a single sample of data, driven to frustration by the mountainous task of unlocking its secrets with no access to other examples. Then, as we began to pick away at the possibilities and impossibilities of his situation, we found, step by step, that he is not so different from the rest of us. The perspective we took consisted of datasets as single bit strings and their analysis by effective methods: computer programs.

So what have we discovered? What secrets can Onno Quist hope to unlock from the Phaistos disc? In Chapter 3, we saw what he can achieve if he is willing to make a model assumption. Under such an assumption, he can compute an approximation and be almost certain that he has approximated the Kolmogorov complexity with good accuracy, at least in an asymptotic sense. This holds for standard model classes, such as DFAs, HMMs or Normal distributions, but the principle also applies to far broader model classes that are not usually explored in statistics, like those defined by a resource bound.

Chapter 4 brought more sobering news. Even if a representation that reaches the Kolmogorov complexity captures all patterns that we can hope to understand, we cannot separate those patterns into structure and noise objectively. The ultimate hope, for Onno, would be to fit a Turing machine to the Phaistos disc, such that running the Turing machine again, with a random input, would cause the machine to produce another example of a Phaistos disc, just as if it had been freshly stamped by the ancient Minoan workshops that produced the original. Here we must disappoint: there are many Turing machines that achieve the Kolmogorov complexity, and the way they respond if we run them with a random input varies greatly. One of them, the universal Turing machine, will produce simply another object, a sample from the universal distribution. This could be anything from a new Shakespeare play, to a proof of the Goldbach conjecture, to completely random noise. Another Turing machine, which compresses the Phaistos disc just as well, when run, will output exactly the same Phaistos disc we know already, every single time. Somewhere in between these two extremes, there may be a solution that does what we had hoped for: another Phaistos disc, like the original one, but not exactly like it. Unfortunately for Onno, without a second sample we have no basis to decide which we should choose. It seems that the choice of which parts of the data are noise and which are structure, can simply not be made objectively.

Still, Onno is not quite forced to make assumptions he can never prove. An alternative strategy, that will let him learn *something* at least, is to make assumptions in the hope of *dis*proving them. Consider for instance, that we have the intuition that successive symbols of the disc should be grouped together, so that each pair forms a "word" in the language of the disc. If this is true, and some such "words" are more likely than others, a compressor that uses this idea may allow us to compress the symbol sequence. Such a result does not prove that the disc consists of words of two symbols, but we can use it to *dis*prove other hypotheses. For instance, if this compressor achieves a shorter codelength than any compressor that casts every individual symbol on the disc as the result of an independent draw from some distribution on the complete alphabet, we have disproved the hypothesis that the disc is simply a sequence of independently drawn symbols from a single distribution.

We use this principle in Chapter 5 to analyze complex graphs. Complex graphs are a relatively new form of dataset, and a lot of traditional techniques do not apply in this domain. A graph does not provide us with easily identifiable "building blocks." In fact, in most complex graphs, every node is a handful of steps removed from every other node, so that even defining the idea of a neighbourhood becomes troublesome. We built on the idea of a *network motif*, a frequently recurring subgraph, and use the principle described above to construct a new, fast method for the discovery of network motifs.

Finally, in Chapter 6, we investigated a more unusual type of structure: self-similarity. Self-similarity occurs if a large part of a dataset is in some sense similar to a scaled down copy of itself. The way a head of broccoli seems to consist of smaller heads of broccoli, the way a sentence may consist of smaller sub-sentences or the way a large organization is organized into smaller sub-organizations. To study such self-similar structure, we made things easy for ourselves and studied the most self similar objects possible: *iterated function systems* (IFSs). In IFSs, the self similarity is exact: they can be cut into parts so that every part is exactly a scaled down copy of the whole. We developed an algorithm that can learn IFS model from a dataset sampled from them, showing that, in principle at least, self-similar structure can be learned.

To establish a thematic connection between the chapters of this thesis, we described a perspective on the business of making inferences from data. We framed a statistician as someone trying to find Turing machines that are likely to generate his data. Data which constitutes a single sample from a computable distribution. While not every working statistician may identify with this view, it has provided us with some insight. We have found that there are things we can't do in general, like consistent model selection, things that only work under highly restricted model-constraints. Other things we saw, remain possible, no matter how broad we make our model class. We have also seen that such a perspective can be highly valuable when a new form of data arrives, such as the complex graph, which defies traditional methods of analysis. In such situations we will see a multitude of methods emerge as each researcher invents new ways of tackling the new form of data, each from his own perspective, tailored to his own needs. How each method compares to the next, how they relate and differ, becomes a difficult question to answer. The single sample setting provides a simple railing against which we can steady ourselves in this storm of new ideas. At the very least, we are all of us faced with a single finite bit string, and we are all hoping to find a computable process that can reproduce the patterns in that string.

References

[1] http://chrisharrison.net/projects/bibleviz/index.html. Accessed: 2014-08-22.

[2] King James network dataset – KONECT, October 2014.

[3] American revolution network dataset – KONECT, April 2015.

[4] Cattle network dataset – KONECT, January 2015.

[5] Physicians network dataset – KONECT, April 2015.

[6] Wikipedia, nl (dynamic) network dataset – KONECT, April 2015.

[7] L. A. Adamic, J. Zhang, E. Bakshy, and M. S. Ackerman. Knowledge sharing and Yahoo answers: everyone knows something. In *Proceedings of the 17th International Conference on World Wide Web, WWW 2008, Beijing, China, April 21-25, 2008*, pages 665–674, 2008.

[8] P. Adriaans. Facticity as the amount of self-descriptive information in a data set. *arXiv preprint arXiv:1203.2245*, 2012.

[9] R. Albert and A. L. Barabási. Statistical mechanics of complex networks. *Reviews of modern physics*, 74(1):47, 2002.

[10] J. E. Angus. Bootstrap one-sided confidence intervals for the log-normal mean. *The Statistician*, pages 395–401, 1994.

[11] L. Antunes, L. Fortnow, D. van Melkebeek, and N. V. Vinodchandran. Computational depth: concept and applications. *Th. Comp. Sc.*, 354(3):391–404, 2006.

[12] L. F. C. Antunes, B. Bauwens, A. Souto, and A. Teixeira. Sophistication vs logical depth, 2013. http://arxiv.org/abs/1304.8046.

[13] L. F. C. Antunes and L. Fortnow. Sophistication revisited. *Theory Comput. Syst.*, 45(1):150–161, 2009.

[14] L. F. C. Antunes, A. Matos, A. Pinto, A. Souto, and A. Teixeira. One-way functions using algorithmic and classical information theories. *Theory Comput. Syst.*, 52(1):162–178,

2013.

[15] L. F. C. Antunes, A. Matos, A. Souto, and P. M. B. Vitányi. Depth as randomness deficiency. *Theory Comput. Syst.*, 45(4):724–739, 2009.

[16] L. Backstrom, P. Boldi, M. Rosa, J. Ugander, and S. Vigna. Four degrees of separation. In *Proceedings of the 4th Annual ACM Web Science Conference*, pages 33–42. ACM, 2012.

[17] M. F. Barnsley. *Fractals everywhere*. Academic press, 2014.

[18] A. Barvinok. On the number of matrices and a random matrix with prescribed row and column sums and 0–1 entries. *Advances in Mathematics*, 224(1):316–339, 2010.

[19] M. J. Beal. *Variational algorithms for approximate Bayesian inference*. University of London, 2003.

[20] C. H. Bennett. Logical depth and physical complexity. In *The Universal Turing Machine: A Half-Century Survey*. Oxford University Press, 1988.

[21] I. Bezáková, A. Kalai, and R. Santhanam. Graph model selection using maximum likelihood. In W. W. Cohen and A. Moore, editors, *Machine Learning, Proceedings of the Twenty-Third International Conference (ICML 2006), Pittsburgh, Pennsylvania, USA, June 25-29, 2006*, volume 148 of *ACM International Conference Proceeding Series*, pages 105–112. ACM, 2006.

[22] J. K. Blitzstein and P. Diaconis. A sequential importance sampling algorithm for generating random graphs with prescribed degrees. *Internet Mathematics*, 6(4):489–522, 2011.

[23] P. Bloem, F. Mota, S. de Rooij, L. Antunes, and P. Adriaans. A safe approximation for Kolmogorov complexity. In *ALT*, pages 336–350, 2014.

[24] C. J. Carstens. Motifs in directed acyclic networks. In *International Conference on Signal-Image Technology & Internet-Based Systems, SITIS 2013, Kyoto, Japan, December 2-5, 2013*, pages 605–611. IEEE, 2013.

[25] J. Chen, W. Hsu, M. Lee, and S. Ng. Nemofinder: dissecting genome-wide protein-protein interactions with meso-scale network motifs. In T. Eliassi-Rad, L. H. Ungar, M. Craven, and D. Gunopulos, editors, *Proceedings of the Twelfth ACM*

SIGKDD International Conference on Knowledge Discovery and Data Mining, Philadelphia, PA, USA, August 20-23, 2006, pages 106–115. ACM, 2006.

[26] Q. Cheng. Multifractal modeling and lacunarity analysis. *Mathematical Geology*, 29(7):919–932, 1997.

[27] R. Cilibrasi and P. M. B. Vitányi. Clustering by compression. *IEEE Transactions on Information Theory*, 51(4):1523–1545, 2005.

[28] J. Coleman, E. Katz, and H. Menzel. The diffusion of an innovation among physicians. *Sociometry*, pages 253–270, 1957.

[29] P. Collet, E. Lutton, F. Raynal, and M. Schoenauer. Polar IFS + Parisian genetic programming ēfficient IFS inverse problem solving. *Genetic Programming and Evolvable Machines*, 1(4):339–361, 2000.

[30] D. J. Cook and L. B. Holder. Substructure discovery using minimum description length and background knowledge. *CoRR*, cs.AI/9402102, 1994.

[31] T. M. Cover. Kolmogorov complexity, data compression, and inference. In *The Impact of Processing Techniques on Communications*, pages 23–33. Springer, 1985.

[32] T. M. Cover and J. A. Thomas. *Elements of information theory (2. ed.)*. Wiley, 2006.

[33] G. Davies. How a statistical formula won the war, July 2006.

[34] R. Deliu, F. Shonkwiler, and A. Mendivil. Genetic algorithms for the 1-d fractal inverse problem. In *Proceedings of the fourth international conference on genetic algorithms*, page 495. Citeseer, 1991.

[35] A. P. Dempster, N. M. Laird, and D. B. Rubin. Maximum likelihood from incomplete data via the em algorithm. *Journal of the royal statistical society. Series B (methodological)*, pages 1–38, 1977.

[36] P. Gács, J. Tromp, and P. M. B. Vitányi. Algorithmic statistics. *IEEE Transactions on Information Theory*, 47(6):2443–2463, 2001.

[37] P. Gács, J. Tromp, and P. M. B. Vitányi. Algorithmic statistics. *IEEE Tr. Inf. Th.*, 47(6):2443–2463, 2001.

[38] J. L. Gailly and M. Adler. The GZIP compressor, 1991.

[39] J. Gehrke, P. Ginsparg, and J. Kleinberg. Overview of the 2003 kdd cup. *ACM SIGKDD Explorations Newsletter*, 5(2):149–151, 2003.

[40] M. Gell-Mann and S. Lloyd. Information measures, effective complexity, and total information. *Complexity*, 2(1):44–52, 1996.

[41] M. Gell-Mann and S. Lloyd. Effective complexity. *Nonextensive Entropy-Interdisciplinary Applications, by Edited by Murray Gell-Mann and C Tsallis, pp. 440. Oxford University Press, Apr 2004. ISBN-10: 0195159764. ISBN-13: 9780195159769*, 1, 2004.

[42] C. I. Del Genio, H. Kim, Z. Toroczkai, and K. E. Bassler. Efficient and exact sampling of simple graphs with given arbitrary degree sequence. *CoRR*, abs/1002.2975, 2010.

[43] E. N. Gilbert. Random graphs. *The Annals of Mathematical Statistics*, pages 1141–1144, 1959.

[44] A. L. Goldberger. Fractal mechanisms in the electrophysiology of the heart. *Engineering in Medicine and Biology Magazine, IEEE*, 11(2):47–52, 1992.

[45] L. Greengard and J. Strain. The fast gauss transform. *SIAM Journal on Scientific and Statistical Computing*, 12(1):79–94, 1991.

[46] P. Grünwald and P. M. B. Vitányi. Shannon information and Kolmogorov complexity, 2004. arXiv:cs/0410002.

[47] P. D. Grünwald. *The Minimum Description Length Principle*. Adaptive computation and machine learning series. The MIT Press, 2007.

[48] J.C. Hart. Fractal image compression and recurrent iterated function systems. *Computer Graphics and Applications, IEEE*, 16(4):25–33, 1996.

[49] J.C. Hart, W.O. Cochran, and P.J. Flynn. Similarity hashing: A computer vision solution to the inverse problem of linear fractals. *FRACTALS-LONDON-*, 5:39–50, 1997.

[50] F. C. Hennie and R. E. Stearns. Two-tape simulation of multitape Turing machines. *J. ACM*, 13(4):533–546, 1966.

[51] J.E. Hutchinson. Deterministic and random fractals. *Complex Systems*, pages 127–166, 2000.

[52] N. Kashtan, S. Itzkovitz, R. Milo, and U. Alon. Efficient sampling algorithm for estimating subgraph concentrations and detecting network motifs. *Bioinformatics*, 20(11):1746–1758, 2004.

[53] J. Kiefer. Sequential minimax search for a maximum. *Proceedings of the American Mathematical Society*, 4(3):502–506, 1953.

[54] H. Kim, C. I. Del Genio, K. E. Bassler, and Z. Toroczkai. Constructing and sampling directed graphs with given degree sequences. *New Journal of Physics*, 14(2):023012, 2012.

[55] S. C. Kleene. On notation for ordinal numbers. *J. Symb. Log.*, pages 150–155, 1938.

[56] A. S. Konagurthu and A. M. Lesk. On the origin of distribution patterns of motifs in biological networks. *BMC Systems Biology*, 2(1):73, 2008.

[57] M. Koppel. Structure. In *The Universal Turing Machine: A Half-Century Survey*. Oxford University Press, 1988.

[58] M. Koppel and H. Atlan. An almost machine-independent theory of program-length complexity, sophistication, and induction. *Inf. Sci.*, 56(1-3):23–33, 1991.

[59] P. S. Laplace. A philosophical essay on probabilities, translated from the 6th french edition by frederick wilson truscott and frederick lincoln emory, 1951.

[60] J. Leskovec, D. Chakrabarti, J. Kleinberg, C. Faloutsos, and Z. Ghahramani. Kronecker graphs: An approach to modeling networks. *The Journal of Machine Learning Research*, 11:985–1042, 2010.

[61] M. Li, X. Chen, X. Li, B. Ma, and P. M. B. Vitányi. The similarity metric. *IEEE Transactions on Information Theory*, 50(12):3250–3264, 2004.

[62] M. Li and P. M. B. Vitányi. *An introduction to Kolmogorov complexity and its applications, Second Edition*. Graduate Texts in Computer Science. Springer, 1997.

[63] M. Li and P.M.B. Vitányi. *An introduction to Kolmogorov complexity and its applications*. Springer-Verlag, 1993.

[64] B. B. Mandelbrot. *The fractal geometry of nature*. Times Books, 1982.

[65] B. B. Mandelbrot. Fractals in physics: squig clusters, dif-

fusions, fractal measures, and the unicity of fractal dimensionality. *Journal of Statistical Physics*, 34(5):895–930, 1984.

[66] B. D. McKay et al. *Practical graph isomorphism*. Department of Computer Science, Vanderbilt University Tennessee, US, 1981.

[67] R. Milo, S. Shen-Orr, S. Itzkovitz, N. Kashtan, D. Chklovskii, and U. Alon. Network motifs: simple building blocks of complex networks. *Science*, 298(5594):824–827, 2002.

[68] F. Mota, S. Aaronson, L. F. C. Antunes, and A. Souto. Sophistication as randomness deficiency. In *DCFS 2013*, pages 172–181, 2013.

[69] H. Mulisch. *The discovery of heaven*. De Bezige Bij, 1992. Personal translation.

[70] A. Myronenko and X. Song. On the closed-form solution of the rotation matrix arising in computer vision problems. *arXiv preprint arXiv:0904.1613*, 2009.

[71] A. Myronenko and X. Song. Point set registration: Coherent point drift. *Pattern Analysis and Machine Intelligence, IEEE Transactions on*, 32(12):2262–2275, 2010.

[72] D.J. Nettleton and R. R. Garigliano. Evolutionary algorithms and a fractal inverse problem. *Biosystems*, 33(3):221–231, 1994.

[73] M. E. J. Newman. *Networks: an introduction*. Oxford University Press, 2010.

[74] J. Preusse, J. Kunegis, M. Thimm, T. Gottron, and S. Staab. Structural dynamics of knowledge networks. In *Proc. Int. Conf. on Weblogs and Social Media*, 2013.

[75] T. Reguly, A. Breitkreutz, L. Boucher, B.J. Breitkreutz, G. C. Hon, C. L. Myers, A. Parsons, H. Friesen, R. Oughtred, A. Tong, et al. Comprehensive curation and analysis of global interaction networks in saccharomyces cerevisiae. *Journal of biology*, 5(4):11, 2006.

[76] A. Renyi and P. Erdős. On random graphs. *Publicationes Mathematicae*, 6(290-297):5, 1959.

[77] P. Ribeiro, F. Silva, and M. Kaiser. Strategies for network motifs discovery. In *e-Science, 2009. e-Science'09. Fifth IEEE International Conference on*, pages 80–87. IEEE, 2009.

[78] R. Rinaldo and A. Zakhor. Inverse and approximation problem for two-dimensional fractal sets. *Image Processing, IEEE Transactions on*, 3(6):802–820, 1994.

[79] J. Rissanen. Modeling by shortest data description. *Automatica*, 14(5):465–471, 1978.

[80] J. Rissanen. Universal coding, information, prediction, and estimation. *IEEE Transactions on Information Theory*, 30(4):629–636, 1984.

[81] J. Rissanen and G. G. Langdon. Arithmetic coding. *IBM Journal of research and development*, 23(2):149–162, 1979.

[82] M. W. Schein and M. H. Fohrman. Social dominance relationships in a herd of dairy cattle. *The British J. of Animal Behaviour*, 3(2):45–55, 1955.

[83] A. Kh. Shen. The concept of (α, β)-stochasticity in the Kolmogorov sense, and its properties. *Soviet Math. Dokl*, 28(1):295–299, 1983.

[84] O. Sporns and R. Kötter. Motifs in brain networks. *PLoS Biol*, 2(11):e369, 2004.

[85] S. A. Terwijn, L. Torenvliet, and P. M. B. Vitányi. Nonapproximability of the normalized information distance. *J. Comput. Syst. Sci.*, 77(4):738–742, 2011.

[86] J. Theiler. Estimating fractal dimension. *JOSA A*, 7(6):1055–1073, 1990.

[87] A. Turiel and C.J. Pérez-Vicente. Multifractal geometry in stock market time series. *Physica A: Statistical Mechanics and its Applications*, 322:629–649, 2003.

[88] A. M. Turing. On computable numbers, with an application to the entscheidungsproblem. *J. of Math*, 58(345-363):5, 1936.

[89] N. Vereshchagin. Algorithmic minimal sufficient statistics: a new approach. *Theory of Computing Systems*, pages 1–19, 2015.

[90] N. K. Vereshchagin and P. M. B. Vitányi. Kolmogorov's structure functions and model selection. *IEEE Tr. Inf. Th.*, 50(12):3265–3290, 2004.

[91] P. M. B. Vitányi. Meaningful information. *IEEE Tr. Inf. Th.*, 52(10), 2004.

[92] P. M. B. Vitányi. Meaningful information. *IEEE Transactions*

on Information Theory, 52(10):4617–4626, 2006.

[93] S. Wernicke. A faster algorithm for detecting network motifs. In R. Casadio and G. Myers, editors, *Algorithms in Bioinformatics, 5th International Workshop, WABI 2005, Mallorca, Spain, October 3-6, 2005, Proceedings*, volume 3692 of *Lecture Notes in Computer Science*, pages 165–177. Springer, 2005.

[94] E. Wong, B. Baur, S. Quader, and C. Huang. Biological network motif detection: principles and practice. *Briefings in bioinformatics*, 13(2):202–215, 2012.

[95] C. Yang, R. Duraiswami, N. Gumerov, L. Davis, et al. Improved fast gauss transform and efficient kernel density estimation. In *Computer Vision, 2003. Proceedings. Ninth IEEE International Conference on*, pages 664–671. IEEE, 2003.

[96] X. H. Zhou and S. Gao. Confidence intervals for the log-normal mean. *Statistics in medicine*, 16(7):783–790, 1997.

Acknowledgements

In computer science writing, the convention is to avoid the singular, first person pronoun "I". Even if a paper has only one author, that author will choose phrases like "we performed the following experiment" and "we consider this...". While this habit probably arose simply because single-author papers are relatively rare, I have always taken it as a tacit acknowledgement that no research is ever completed without the help of others, even when only one person claims authorship. That is never more true than in the case of a dissertation, and this one is no exception.

From observing both my own PhD process, and those of others, I have come to appreciate the value of a good supervisor. I have seen how bad things can get when supervision is lacking, and how much is added when when things go right. In this respect I have been extraordinarily fortunate.

Getting a PhD position in the first place was a difficult process, and I am certain that without Pieter Adriaans, I would never have had the opportunity to work on subjects like Kolmogorov complexity, complex graphs and fractals. The very subjects that topped my wishlist when I was looking for positions four years ago. Since then, Pieter has been a never-ending source of inspiration, never afraid to draw connections between computer science and physics, philosophy and even art. I have never been one to enjoy computer science for its own sake, and it's these connections that have kept me interested. Even if the answers we found were sometimes negative, the questions driving our research were always the right ones.

To Steven de Rooij, my co-promotor, I probably owe this PhD. When I started, like many PhD students I felt like I had done enough learning, and nobody needed to show me any ropes. This translated to far too long a period without publications, and when he joined our group Steven had the thankless job of showing me that I had plenty left to learn. Both about the business of doing science, and about the business of communicating it.

Which is not to say that our interactions were unpleasant. Apart from the occasional clash of a critical mind with a stubborn one, I have enjoyed not only the way we worked together, but also the way we didn't, indulging in long conversations, ranging from pop-culture to gender politics. It may have been detrimental to my productivity, but it was certainly beneficial to my worldview.

Working with Gerben and Wibi, the two other members of our little clique, was never less than a pleasure. The same goes for the rest of the SNE group. I can only apologize for showering you with formulas group meeting after group meeting. I am certain that I will remember my time at the UvA fondly, and that is down to you.

One of the greater challenges of my PhD was learning to do the truly theoretical work shown in the first two chapters of this thesis. Before I started I had proved to myself that I could *understand* such material, but I wasn't at all sure that I could *produce* it. The first signs that things weren't entirely hopeless came during two months in Portugal, working with Luís Antunes and Francisco Mota. The opportunity they gave me to focus for an extended time on a single problem, coupled with the incredible beauty of Porto in spring time, made those first steps that much easier.

The last year, more than any, has been one of single-minded focus. I must thank Gijs, Sandrijn, Danny, Sophie and Natalia, not just for being there, but also for putting up with me not being there. The less we met, the more it meant, and now that the thing is done, I hope we can make up for lost time.

Finally, I want to thank my parents, Kees and Els. As I said, most of my motivation to do this sort of thing derives from an interest not just in science, but in the areas where it intersects with subjects like philosophy and art. I can trace those influences back to my childhood and the environment I grew up in. So I can think of no two people more deserving of my acknowledgement.

Dutch summary

Wat kunnen we leren van een enkel voorbeeld? Als een ingewik-
keld proces grote, complexe objecten genereert, en we krijgen
maar één voorbeeld van zo'n object, kunnen we dan nog conclu-
sies trekken over het proces? Als we patronen vinden, kunnen we
daar een betekenis aan toekennen? Dit is niet alleen een acade-
mische vraag. Er is maar één Internet, bijvoorbeeld, en maar één
wereldwijd financieel systeem. Wat voor soort aannames moeten we doen over het proces dat dit "object" genereerde, om iets
over hun structuur te leren? Wat kunnen we voor elkaar krijgen
als we helemaal niks aannemen? Ieder hoofdstuk in deze disser-
tatie onderzoekt een aspect van deze vraag, beginnend met een
theoretische blik, die stap voor stap praktischer wordt.

De eerste twee hoofdstukken bieden een informele introduc-
tie tot de onderwerpen die steeds terugkeren. In Hoofdstuk 3
onderzoeken we het probleem in zijn meest generieke vorm, met
de theorie van *Kolmogorov complexiteit*. Kolmogorov complexiteit
is een manier om data te analyseren met maar één aanname: dat
de bron van de data als berekenbaar proces gezien kan worden.
Onder deze aanname biedt Kolmogorov complexiteit een formele
definitie van de hoeveelheid informatie die de data bevat. De Kolmogorov complexiteit zelf is niet berekenbaar, maar er zijn bere-
kenbare functies die als bovengrens kunnen dienen. Door aana-
mes te doen over de bron van de data kunnen we een approximatie van de Kolmogorov complexiteit bereken die met hoge kans
dicht bij de daadwerkelijke waarde ligt. We analyseren ook functies die afgeleid zijn van de Kolmogorov complexiteit. We laten
zien dat met een goede approximatie van de Kolmogorov complexiteit, we nog niet per se direct een goede approximatie van
de afgeleide functies hebben, maar met een zorgvuldige analyse
is het mogelijk om bepaalde garanties te bieden.

Hoofdstuk 4 gaat over modelselectie. Als we slechts een enkel
voorbeeld hebben kunnen we dan iets zeggen over de complexiteit van de bron van de data? Hoeveel van de data is structuur en

hoeveel is ruis? De studie van deze vraag heeft verschillende namen: *sophistication*, de *algorithmic sufficient statistic* en *effective complexity*. We laten zien dat al deze aanpakken fundamentele problemen hebben. De voorgestelde functies kunnen niet corresponderen met de intuïtie die ze motiveerde. Het blijft een open vraag of objectieve modelselectie in deze zin mogelijk is, maar we geven verschillende redenen om te geloven dat dit niet zo is.

In Hoofdstuk 5 behandelen we een praktisch probleem: de analyse van grote complexe grafen. Dit zijn complexe objecten, rijk aan interne structuur, maar zonder voor-de-hand-liggende manier om de data in stukken te verdelen die op elkaar lijken. Een populaire methode is om te zoeken naar kleine, veel voorkomende subgrafen: *network motifs*. Het feit dat een subgraaf vaak voorkomt is echter niet op zichzelf een indicatie dat het ook een betekenisvol patroon is: veel subgrafen komen simpelweg vaak voor in iedere willekeurige graaf. Om te laten zien dat zo'n subgraaf belangrijk is, moeten laten zien dat hij *onverwacht* vaak voorkomt. We maken gebruik van het *Minimum Description Length* principe, het praktische broertje van de Kolmogorov complexiteit, om een nieuwe methode te ontwikkeling waarmee we snel kunnen aantonen dat de frequentie van een subgraaf onverwacht hoog is. Hiermee kunnen we dit soort analyses opschalen naar veel grotere grafen dan tot nu toe mogelijk was.

Waar het laatste hoofdstuk terugkerende structuren op dezelfde schaal behandelde, gaat Hoofdstuk 6 in op *self-similarity*; terugkerende structuren op verschillende schalen. Dit is vaak een cruciale aanname in graafanalyse. We kunnen bijvoorbeeld niet het hele *world wide web* analyseren, dus nemen we aan dat een kleine subset dezelfde eigenschappen als het hele web heeft. Het leren van dit soort structuur is het *fractal inverse* probleem, een belangrijke open vraag. We analyseren dit probleem in termen van distributies in Euclidische ruimten, en we laten zien dat het met een EM algoritme aan te pakken is.

Het vakgebied van statistiek is netjes opgedeeld naar datatype. Voor iedere soort data bestaat een aparte familie van technieken. Het voorbeeld van statistiek op een enkel voorbeeld biedt ons een overkoepelend perspectief: in alle gevallen analyseren we feitelijk een enkele bitstring uit een berekenbaar proces. Het "type" van de data is simpelweg een aanname die we doen over de bron, vaak zodat we de data in stukken kunnen knippen, op zo'n manier dat de overeenkomsten tussen die stukken informatie bieden over de bron van de data. Dit perspectief wordt nuttig als we moderne vormen van data, zoals complexe grafen, tegenkomen, en de vraag hoe de data in stukken geknipt kan worden niet makkelijk meer te beantwoorden is. In die situatie kunnen we de data altijd opvatten als een bitstring, gegenereerd door een berekenbaar proces.

A · PROOFS AND DERIVATIONS

Chapter 3

A.0.1 TMs and lsc. Probability Semimeasures (Lemma 3.1)

Definition A.1. A function $f : \mathbb{B} \to \mathbb{R}$ is *lower semicomputable* *(lsc.)* iff there exists a total, computable two-argument function $f' : \mathbb{B} \times \mathbb{N} \to \mathbb{Q}$ such that: $\lim_{i \to \infty} f'(x, i) = f(x)$ and for all i, $f'(x, i+1) \geqslant f'(x, i)$.

Lemma A.1. If f is an lsc. probability semimeasure, then there exists a a function $f^*(x, i)$ with the same properties of the function f' from Definition A.1, and the additional property that all values returned by f^* have finite binary expansions.

Proof. Let x_j represent $x \in \mathbb{D}$ truncated at the first j bits of its binary expansion and x^j the remainder. Let $f^*(x, i) = f'(x, i)_i$. Since $f'(x, i) - f^*(x, i)_i$ is a value with $i+1$ as the highest non-zero bit in its binary expansion, $\lim_{i \to \infty} f^*(x, i) = \lim f'(x, i) = f(x)$.

It remains to show that f^* is nondecreasing in i. Let $x \geqslant y$. We will show that $x_j \geqslant y_j$, and thus $x_{j+1} \geqslant y_j$. If $x = y$ the result follows trivially. Otherwise, we have $x_j = x - x^j > y - x^j = y_j + y^j - x^j \geqslant y_j - 2^{-j}$. Substituting $x = f'(x, i+1)$ and $y = f'(x, i)$ tells us that $f^*(x, i+1) \geqslant f^*(x, i)$ □

Theorem A.1. Any TM, T_q, samples from an lsc. probability semimeasure.

Proof. We will define a program computing a function $p'_q(x, i)$ to approximate $p_q(x)$: Dovetail the computation of T_q on all inputs $x \in \mathbb{B}$ for i cycles.

Clearly this function is nondecreasing. To show that it goes to $p(x)$ with i, we first note that for a given i_0 there is a j such that, $2^{-j-1} < p_q(x) - p_q(x, i_0) \leqslant 2^{-j}$. Let $\{p_i\}$ be an ordering of the programs producing x, by increasing length, that have not yet stopped at dovetailing cycle i_0. There is an m such that $\sum_{i=1}^{m} 2^{-|p_i|} \geqslant 2^{-j-1}$, since $\sum_{i=1}^{\infty} 2^{-|p_i|} > 2^{-j-i}$. Let i_1 be the dovetailing cycle for which the last program below p_{m+1} halts. This gives us $p_q(x) - p_q(x, i_1) \leqslant 2^{-j-1}$. Thus, by induction, we can choose i to make $p(x) - p'(x, i)$ arbitrarily small.   □

Theorem A.2. Any lsc. probability semimeasure can be sampled by a TM.

Proof. Let $p(x)$ be an lsc. probability semimeasure and $p^*(x, i)$ as in Lemma A.1. We assume—without loss of generality—that $p^*(x, 0) = 0$. Consider the following algorithm:

> **initialize** $s \leftarrow \epsilon, r \leftarrow \epsilon$
> **for** $c = 1, 2, \ldots$:
>> **for** $x \in \{b \in \mathbb{B} : |b| \leqslant c\}$
>>> $d \leftarrow p^*(x, c - i + 1) - p^*(x, c - i)$
>>> $s \leftarrow s + d$
>>> add a random bit to r until it is as long as s
>>> **if** $r < s$ **then return** x

The reader may verify that this program dovetails computation of $p^*(x, i)$ for increasing i for all x; the variable s contains the summed probability mass that has been encountered so far. Whenever s is incremented, mentally associate the interval $(s, s + d]$ with outcome x. Since $p^*(x, i)$ goes to $p(x)$ as i increases, the summed length of the intervals associated with x goes to $p(x)$ and s itself goes to $\bar{s} = \sum_x p(x)$. We can therefore sample from p by picking a number r that is uniformly random on $[0, 1]$ and returning the outcome associated with the interval containing r. Since s must have finite length (due to the construction of p^*), we only need to know r up to finite precision to be able to determine which interval it falls in; this allows us to generate r on the fly. Theprogram halts unless r falls in the interval $[\bar{s}, 1]$, which corresponds exactly to the deficiency of p: if p is a semimeasure, we expect the non-halting probability of a TM sampling it to correspond to $1 - \sum_x p(x)$.   □

Theorems A.1 and A.2 combined prove that the class of distributions sampled by Turing machines equals the lower semicomputable semimeasures (Lemma 3.1).

A.0.2 Domination of model class supersets

Lemma A.2. Let C and D be model classes. If $C \subseteq D$, then m^D dominates m^C:

$$\frac{m^D(x)}{m^C(x)} \geqslant \alpha$$

for some constant α independent of x.

Proof. We can partition the models of D into those belonging to C and the rest, which we will call $\bar{C}$. For any given enumeration of D, we get $m^D(x) = \alpha m^C(x) + (1 - \alpha)m^{\bar{C}(x)}$. This gives us:

$$\frac{m^D(x)}{m^C(x)} = \alpha + (1 - \alpha)\frac{m^{\bar{C}(x)}}{m^C(x)} \geqslant \alpha.$$

$\square$

A.0.3 Unsafe Approximation of ID (Theorem 3.5)

Theorem A.3. Under the following assumptions:

- C contains a model T_0, with $p_0(x) = 2^{-|x|}s(|x|)$, with s a distribution on $\mathbb{N}$ which decays polynomially or slower,
- there exists a model-bounded one-way function f for C,
- C is *normal*, i.e. for some c and all x: $\kappa^C(x) < |\bar{x}| + c$

ID^C is an unsafe approximation for ID against an adversary T_q which samples x from p_0 and returns $\bar{x}f(x)$.

Proof.

$$p_q\left(ID^C(x, y) - ID(x, y) \geqslant k\right) =$$

$$p_0\left(\max\left[\bar{\kappa}^C(x \mid f(x)), \bar{\kappa}^C(f(x) \mid x)\right] - \max\left[K(x \mid f(x)), K(f(x) \mid x)\right] \geqslant k\right).$$

$$p_q\left(|x| - ID^C(x, y) \geqslant 2k\right) \leqslant p_0\left(|x| - \bar{\kappa}^C(x \mid f(x)) \geqslant 2k\right)$$

$$\leqslant p_0\left(|x| - \kappa^C(x) \geqslant k \vee \kappa^C(x) - \bar{\kappa}^C(x \mid f(x)) \geqslant k\right)$$

$$\leqslant p_0\left(|x| - \kappa^C(x) \geqslant k\right) + p_0\left(\kappa^C(x) - \kappa^C(x \mid f(x)) \geqslant k\right)$$

$$\leqslant 2^{-k} + cb^{-k}.$$

K can invert $f(x)$, so

$$ID(x, y) = \max \left[K(x \mid f(x)), K(f(x) \mid x) \right] = \max \left[|f^*|, |f^*_{inv}| \right] < c_f$$

where f^* and f^*_{inv} are the shortest program to compute f on U and the shortest program to compute the inverse of f on U respectively.

$$p_q \left(ID^C(x, y) - ID(x, y) \geqslant k \right) + p_q \left(|x| - ID^C(x, y) \geqslant k \right)$$

$$\geqslant p_q \left(ID^C(x, y) - ID(x, y) \geqslant k \vee |x| - ID^C(x, y) \geqslant k \right)$$

$$\geqslant p_q \left(|x| - ID(x, y) \geqslant k \right) \geqslant p_0 \left(|x| - c_f \geqslant k \right)$$

$$= \sum_{i \geqslant k - c_f} s(i) .$$

Which gives us:

$$p_q \left(ID^C(x, y) - ID(x, y) \geqslant k \right)$$

$$\geqslant -p_q(|x| - ID^C \geqslant k) + \sum_{i \geqslant k - |f|} s(i)$$

$$\geqslant -cb^{-k} + \sum_{i \geqslant k - |f|} s(i)$$

$$\geqslant s(k - |f|) - cb^{-k} \geqslant c's(k) \qquad \text{for the right } c'. \qquad \square$$

Corollary A.1. Under the assumptions of Theorem 3.5, $\overline{\kappa}^C(x \mid y)$ is an unsafe approximation for $K(x \mid y)$ against q.

Proof. Assuming $\overline{\kappa}^C$ is safe, then since max is safety-preserving (Lemma A.4), ID^C should be safe for ID. Since it isn't, $\overline{\kappa}^C$ cannot be safe. $\qquad \square$

A.0.4 Safe Approximation of ID (Theorem 3.6)

Lemma A.3. If q samples x and y independently from models in C, then $\kappa^C(x \mid y)$ is a 2-safe approximation of $-\log m(x \mid y)$ against q.

Proof. Let q sample x from p_r and y from p_s.

$$p_q(-\log m^C(x \mid y) + \log m(x \mid y) \geqslant k)$$

$$= p_q(m(x \mid y)/m^C(x \mid y) \geqslant 2^k)$$

$$\leqslant 2^{-k}E\left[m(x \mid y)/m^C(x \mid y)\right]$$

$$= 2^{-k}\sum_{x,y} p_s(y)m(x \mid y)\frac{p_r(x)}{m^C(x \mid y)}$$

$$\leqslant c2^{-k}\sum_{x,y} p_s(y)m(x \mid y)\frac{m^C(x \mid y)}{m^C(x \mid y)}$$

$$\leqslant c2^{-k}\sum_{x,y} p_s(y)m(x \mid y) \leqslant c2^{-k}. \qquad \square$$

Since m and K mutually dominate, $-\log m^C$ is 2-safe for $K(x \mid y)$, as is $\overline{K}(x \mid y)$.

Lemma A.4. If f_a is safe for f against q, and g_a is safe for g against q, then $\max(f_a, g_a)$ is safe for $\max(f, g)$ against q.[1]

Proof. We first partition $\mathbb{B}$ into sets A_k and B_k:

$A_k = \{x : f_a(x) - f(x) \geqslant k \lor g_a - g(x) \geqslant k\}$ Since both f_a and g_a are safe, we know that $p_q(A_k)$ will be bounded above by the sum of two inverse exponentials in k, which from a given k_0 is itself bounded by an exponential in k.

$B_k = \{x : f_a(x) - f(x) < k \land g_a - g(x) < k\}$ We want to show that B contains no strings with error over k. If, for a given x the left and right max functions in $\max(f_a, g_a) - \max(f, g)$ select the outcome from matching functions, and the error is below k by definition. Assume then, that a different function is selected on each side. Without loss of generality, we can say that $\max(f_a, g_a) = f_a$ and $\max(f, g) = g$. This gives us: $\max(f_a, g_a) - \max(f, g) = f_a - g \leqslant f_a - f \leqslant k$.

We now have $p(B_k) = 0$ and $p(A_k) \leqslant cb^{-k}$, from which the theorem follows. $\qquad \square$

---

[1] We will call such operations *safety preserving*

Corollary A.2. ID^C is a safe approximation of ID against sources that sample x and y independently from models in C.

A.0.5 Safe approximation of NID (Theorem 3.7)

Lemma A.5. Let f and g be two functions, with f_a and g_a their safe approximations against adversary p_q. Let $h(x) = f(x)/g(x)$ and $h_a(x) = f_a(x)/g_a(x)$. Let $c > 1$ and $0 < \epsilon \ll 1$ be constants such that $p_q(f_a(x) \geqslant c) \leqslant \epsilon$ and $p_q(g_a(x) \geqslant c) \leqslant \epsilon$. We can show that for some $b > 1$ and $c > 0$

$$p_q\left(\left|\frac{h(x)}{h_a(x)} - 1\right| \geqslant \frac{k}{c}\right) \leqslant cb^{-k} + 2\epsilon.$$

Proof. We will first prove the bound from above, using f_a's safety, and then the bound from below using g_a's safety.

$$p_q\left(\frac{h}{h_a} \leqslant 1 - \frac{k}{c}\right) \leqslant p_q\left(\frac{h}{h_a} \leqslant 1 - \frac{k}{c} \ \& \ c < f_a\right) + \epsilon$$

$$\leqslant p_q\left(\frac{h}{h_a} \leqslant 1 - \frac{k}{f_a}\right) + \epsilon$$

$$= p_q\left(\frac{f}{f_a}\frac{g_a}{g} \leqslant 1 - \frac{k}{f_a}\right) + \epsilon \leqslant p_q\left(\frac{f}{f_a} \leqslant 1 - \frac{k}{f_a}\right) + \epsilon$$

$$= p_q\left(\frac{f+k}{f_a} \leqslant 1\right) + \epsilon = p_q\left(f_a - f \geqslant k\right) + \epsilon \leqslant c_f b_f^{-k} + \epsilon.$$

The other bound we prove similarly. Combining the two, we get

$$p_q\left(h/h_a \notin (k/c - 1, k/c + 1)\right) \leqslant c_f b_f^{-k} + c_g b_g^{-k} + 2\epsilon$$

$$\leqslant c'b'^{-k} + 2\epsilon. \qquad \square$$

Theorem 3.7 follows as a corollary.

Chapter 4

Lemma A.6 (Invariance of function complexity). Let ψ and η be any two acceptable numberings Let f be any partial computable function. There exists a constant c independent of f such that

$$\left|C^{\mathcal{K},\psi}(f) - C^{\mathcal{K},\eta}(f)\right| \leqslant c \text{ and } \left|C^{\mathcal{C},\psi}(f) - C^{\mathcal{C},\eta}(f)\right| \leqslant c.$$

Proof. Let $g(i)$ be the function such that $\psi_i = \eta_{g(i)}$.

$$C^{\mathcal{C},\psi}(f) = \min\left\{C^{\mathcal{C},\psi}(i) : \psi_i = f\right\} \geqslant \min\left\{C^{\mathcal{C},\eta}(i) : \psi_i = f\right\} - c$$

$$= \min\left\{C^{\mathcal{C},\eta}(i) : \eta_{g(i)} = f\right\} - c$$

$$= \min\left\{C^{\mathcal{C},\eta}(g(i)) : \eta_{g(i)} = f\right\} - c'$$

$$\geqslant \min\left\{C^{\mathcal{C},\eta}(j) : \eta_j = f\right\} - c' = C^{\mathcal{C},\eta}(f).$$

Reverse ψ and η for the opposite inequality. The same proof holds for $C^{\mathcal{K}}$. $\qquad\square$

Chapter 5

Confidence Intervals for the Degree-Sequence Model   As mentioned in the body of the text, even with the highly optimized implementations described in [42] and [54] sampling can be slow for large graphs. In our implementation, a modern day laptop can take several minutes to produce a single sample for a random graph with 10^4 nodes and 10^5 links. However, we are not interested in precision beyond several orders of magnitude, so if we have a reliable method for determining confidence intervals, we can use those to provide us with safe bounds. Since we are dealing with a log-normal source, we cannot simply use twice the standard error of the mean to approximate our error bars. We will use the parametric bootstrap procedure provided in [10, 96]. To substantiate this method, we test the coverage of the two-sided symmetric confidence interval on three datasets. We proceed as follows: first we estimate the true value of $|\mathcal{G}_D|$ with the ML estimator, using 10^6 samples. Call this value g. We use this as our gold standard. We then sample a small number (5, 10, 20) of graphs and their associated probabilities. Using the bootstrap method we construct a two-sided symmetric confidence interval with $\alpha = 0.05$ on this sample. We repeat the procedure of sampling data and constructing an interval 5000 times and report the proportion of times g was inside the confidence interval. If the bootstrap method is accurate, the resulting value should be close to 0.95. We use the following datasets:

	5 samples		10 samples		20 samples	
cattle	0.93	511 (379)	0.94	194 (94.6)	0.94	107 (35.3)
revolution	0.89	147 (120)	0.92	59.3 (29.2)	0.93	33.5 (11.3)
random	0.91	108 (97.0)	0.94	44.7 (24.5)	0.94	25.3 (9.44)

Table A.1: Coverage and interval length at $\alpha = 0.05$ for three small datasets. The coverage is the number of times the true value g was contained in the interval, over 5000 trials. The length is the mean length of these intervals in bits. The standard deviation is given in parentheses.

cattle Observed dominance behaviors between cows. A directed graph with 28 nodes, and 217 links. [82, 4]

revolution Affiliations of 136 people to 5 organizations encoded as a bipartite graph. 141 nodes and 160 links. [3]

random A simple undirected random graph of 50 nodes, with each pair of distinct nodes connected with probability 0.5.

As Table A.1 shows, this method becomes relatively reliable at around 10 samples, although the intervals are quite large at that sample size.

Now, when we use L^{DS} as a base model in L^{motif}, the intractable value $|\mathcal{G}_D|$ occurs in two places: the encoding of the subgraph, and the encoding of the template graph. Since we use our estimator for both, we must be careful to end up with a correct confidence interval for the resulting motif code. Let D' be the degree sequence of the subgraph, and D be the degree sequence of the template. Then, we can split the total codelength into three components: $\log|\mathcal{G}_{D'}|$, $\log|\mathcal{G}_D|$ and R. R is the sum of all parts of the code that we can compute exactly, including the sizes and sequences of the motif and template graph (i.e. everything but $\log|\mathcal{G}_{D'}|$ and $\log|\mathcal{G}_D|$). The total codelength is described by $\log|\mathcal{G}_{D'}|+\log|\mathcal{G}_D|+R$, where the first two terms require the use of the estimator. Let Q_m and Q_h be random variables representing the inverse probability of graphs sampled from the importance sampling algorithm, for the degree sequence of the motif and the template graph respectively. In other words, the true codelength for the motif code is:

$$\log EQ_m + \log EQ_h + R$$

$$= \log EQ_m EQ_h + R$$

$$= \log E[Q_m Q_h] + R$$

where the last line follows from the fact Q_m and Q_h are independent. So to get a correct confidence interval, we can take the same number of samples of both Q_m and Q_h, multiply their probabilities, and perform the bootstrap analysis on the list of these multiplied probabilities (since we are summing the logarithms of log-normally distributed variables, the result is log-normally distributed as well).

Chapter 6

A.0.6 Derivations

Transforming a generic spherical MVN by a similitude can be cast as the transformation of $\mathcal{N}_0$ by two similitudes:

$$f_{t,R,s}(\mathcal{N}_{t_0,s_0^2 I})(x)$$

$$= f_{t,R,s}(f_{t_0,R_0,s_0}(\mathcal{N}_0))(x) = f_{sR t_0+t,RR_0,ss_0}(\mathcal{N}_0)(x)$$

$$= (ss_0)^{-H}\mathcal{N}_0 \left(\frac{1}{ss_0}R_0^{\mathsf{T}}R^{\mathsf{T}}x - \frac{s}{ss_0}R_0^{\mathsf{T}}t_0 - \frac{1}{ss_0}R_0^{\mathsf{T}}R^{\mathsf{T}}t \right)$$

$$= (ss_0)^{-H}\mathcal{N}_0 \left(\frac{1}{ss_0}x - \frac{1}{s_0}Rt_0 - \frac{1}{ss_0}t \right)$$

$$= (ss_0)^{-H}\mathcal{N}_0 \left(\frac{1}{ss_0}R_0^{\mathsf{T}}R^{\mathsf{T}}x - \frac{s}{ss_0}R_0^{\mathsf{T}}t_0 - \frac{1}{ss_0}R_0^{\mathsf{T}}R^{\mathsf{T}}t \right)$$

$$= (2\pi)^{-\frac{H}{2}}(ss_0)^{-H} \exp\left[-\frac{1}{2s_0^2 s^2} \|x - t - sRt_0\|^2 \right]$$

The depth v and weights w   For the depth priors v, the Q-function, with constant terms omitted, reduces to:

$$Q(v) = \sum_{i,j} P_{ij} \ln v_{|c_j|}$$

$$= \sum_{d \in [0,D]} \sum_{j:|c_j|=d} P_{ij} \ln v_d = \sum_{d \in [0,D]} \left[\sum_{j:|c_j|=d} P_{ij} \right] \ln v_d$$

$$= \sum_{d \in [0,D]} p_d \ln v_d \text{ with } p_d = \sum_{j:|c_j|=d} P_{ij}.$$

For v, we have the additional constraint that $\sum_d v_d = 1$, which we can take into consideration with Lagrange multipliers, giving us the objective function $\mathcal{L}(v,\lambda) = p_d/v_d - \lambda \sum_d v_d - \lambda$. We differentiate and set equal to zero, giving us the equations $p_d/v_d - \lambda = 0$ amd $\sum_i v_i - 1 = 0$. Solving these gives us:

$$\hat{v}_d = \frac{p_d}{\sum_i p_i}.$$

Finding w_k follows the same principle as the depths. The Q-function reduces to:

$$Q(w) = \sum_k p_k \ln w_k.$$

Using Lagrange multipliers to incorporate the constraint that $\sum_i w_i = 1$, we find

$$\hat{w}_k = p^k / \sum_i p^i.$$

The translations t_k The optimal translation can be found by straightforward differentiation:

$$\frac{\partial Q(t_k)}{\partial t_k} = -\sum_{i,j} P_{ij}^k \frac{1}{s_j^2 s_k^2} \frac{\|y_i - t_k - s_k R_k t_j\|^2}{\partial t_k}$$

$$= \sum_{i,j} P_{ij}^k \frac{1}{s_j^2 s_k^2} (y_i - t_k - s_k R_k t_j)^\top .$$

This sum represents a row vector, which we transpose to get a column vector, and set equal to zero to get $\hat{t}_k$:

$$0 = \frac{1}{s_k^2} \left(\sum_{i,j} P_{ij}^k s_j^{-2} y_i - \sum_{i,j} P_{ij}^k s_j^{-2} t_k - s_k R_k \sum_{i,j} P_{ij}^k s_j^{-2} t_j \right) .$$

Let $Z = \text{diag}(s_1^{-2}, \ldots, s_M^{-2})$, we can then rewrite to matrix notation:

$$\hat{t}_k = \frac{Y P^k Z 1 - s_k R_k \, T (P^k Z)^\top 1}{1^\top P^k Z 1} .$$

The rotations R_k

$$Q(R_k) = -\frac{1}{2} \sum_{i,j} P_{ij}^k \frac{1}{s_j^2 s_k^2} \|y_i - t_k - s_k R_k t_j\|^2$$

$$= -\frac{1}{2} \sum_{i,j} P_{ij}^k s_j^{-2} \left\| (y_i - y^k) - s_k R_k (t_j - t^k) \right\|^2 .$$

Let $y_i^k = y_i - y^k$ and $t_j^k = t_j - t^k$, i.e. mean-centered versions of the data points and the component means. Let Y^k and T^k be the corresponding matrices with these vectors as columns. We multiply out the inner product to get

$$Q(R_k) = -\frac{1}{2} \sum_{i,j} P^k_{ij} s_j^{-2} \left[y_i^{k^T} y_i^k - 2 s_k y_i^{k^T} R_k t_j^k + t_j^{k^T} t_j^k \right]$$

from which we remove the terms and global factors that are independent of R_k:

$$Q(R_k) = \sum_{i,j} P^k_{ij} \, s_j^{-2} \, y_i^{k^T} R_k t_j^k .$$

Which in matrix notation becomes:

$$Q(R_k) = \mathrm{tr}((P^k Z)^T Y^{k^T} R_k T^k) = \mathrm{tr}(T^k Z P^{k^T} Y^{k^T} R_k)$$

$$= \mathrm{tr}([Y^k P^k Z T^{k^T}]^T R_k) .$$

Where we use the fact that $\mathrm{tr}(A^T B) = \sum_{i,j} (A \circ B)_{ij}$ and the fact that the trace is invariant under cyclic permutations.

The scaling s_k We derive the scaling by filling in $\hat{t}_k$ and solving for $\partial Q(s_k)/\partial s_k$.

$$\frac{\partial Q(s_k)}{\partial s_k} = \frac{-p^k H \ln s_k - \frac{1}{2} \sum_{i,j} (P^k Z)_{ij} s_k^{-2} \| y_i - t_k - s_k R_k t_j \|^2}{\partial s_k}$$

$$\frac{\partial Q(s_k)}{\partial s_k} = -p^k H s_k^{-1} - \frac{1}{2} \sum_{i,j} (P^k Z)_{ij} \left(-2 s_k^{-3} y_i^{k^T} y_i^k + 2 s_k^{-2} y_i^{k^T} R_k t_j^k \right)$$

$$= -p^k H s_k^{-1} + s_k^{-3} \sum_{i,j} (P^k Z)_{ij} s_k^{-3} y_i^{k^T} y_i^k$$

$$- s_k^{-1} \sum_{i,j} (P^k Z)_{ij} s_k^{-2} y_i^{k^T} R_k t_j^k .$$

In matrix notation:

$$0 = s_k^{-2} \mathrm{tr}(\overline{Y}^{k^T} d(P^k Z 1) Y^k) + s_k^{-1} \mathrm{tr}(T^k Z P^{k^T} T^{k^T} R_k) - p^k H .$$

B · Fractal experiments: full results

In Chapter 6, we showed that our algorithm could construct fractal models for various datasets. However, we selected the best result from 100 trials. To give an indication of how likely the algorithm is to converge to such models, we provide all 100 results for all trials. For each trial, we show a single image that represents the model after 300 iterations.

cloud

coast

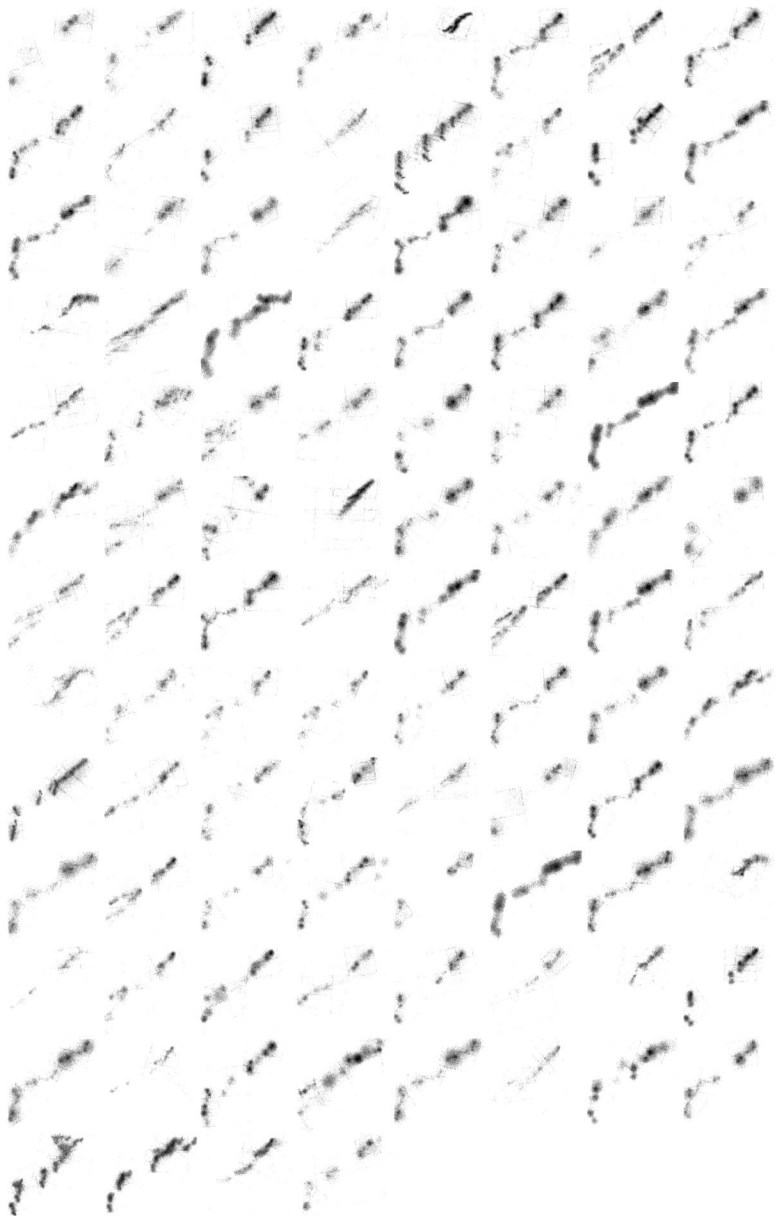

disc

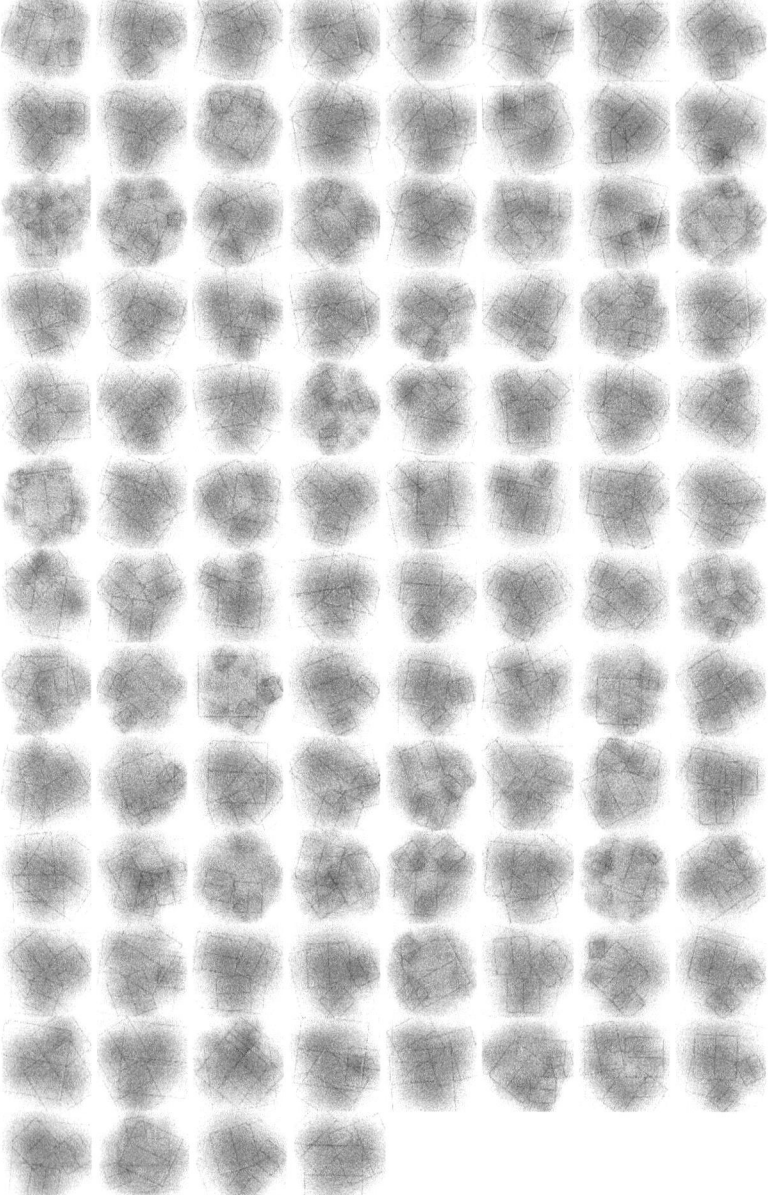

koch2

koch4

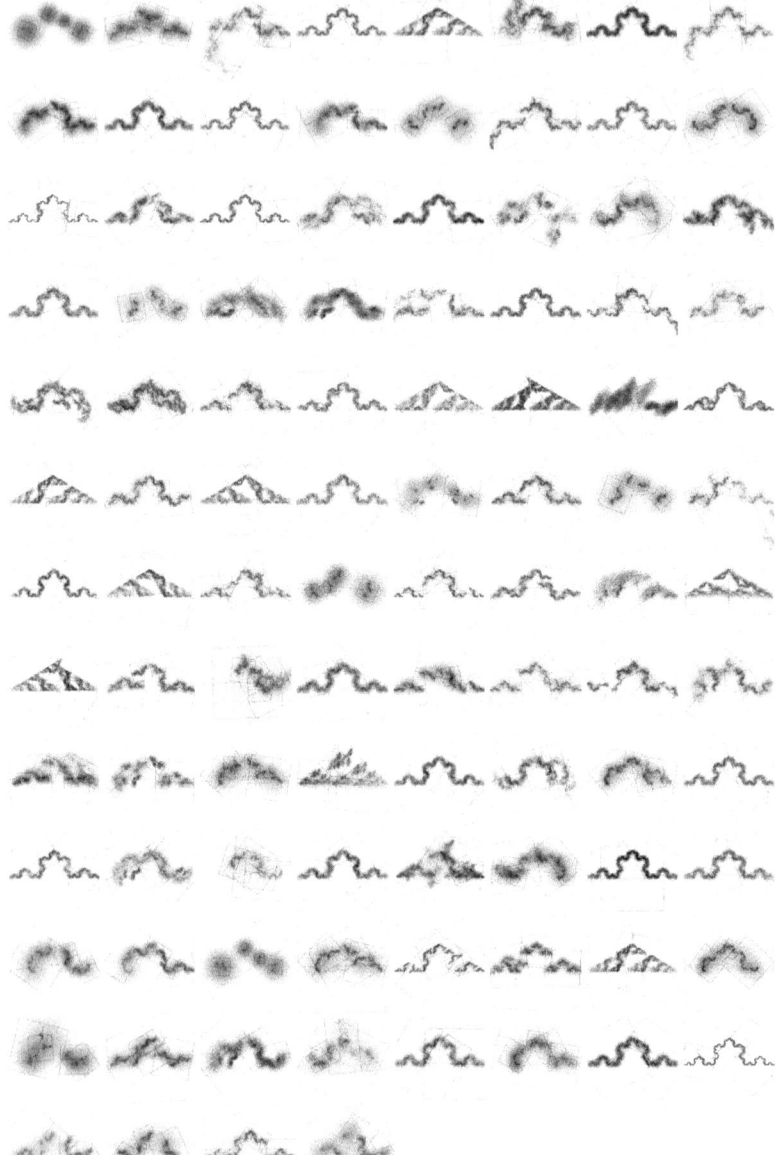

romanesco

sierpinski

sierpinski, unequal weights

sierpinski, with 2 components

sphere

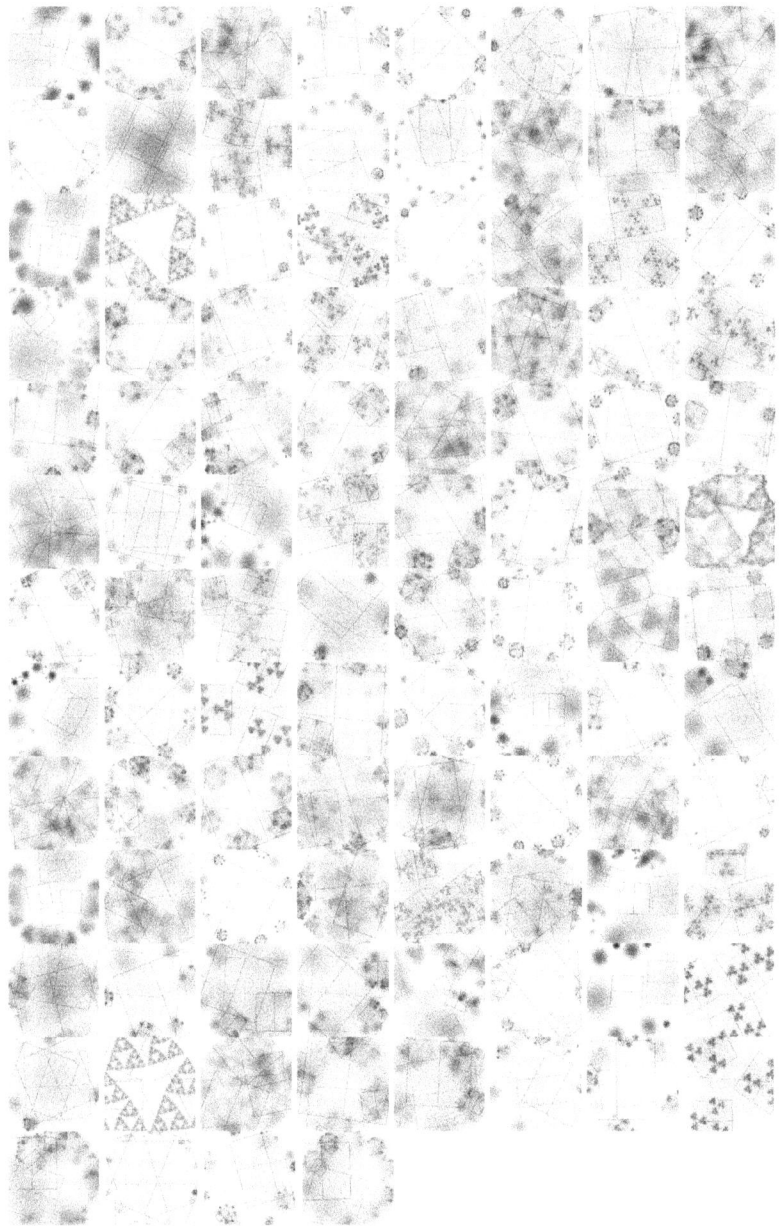

CHANGES MADE AFTER ACCEPTANCE

The following non-cosmetic changes were made to the manuscript after acceptance by the committee:

- On page 4 a claim was corrected. The original pargraph suggested that no inference is possible on models outside the class of Turing machines.
- On page 37, the motivation for the definition of safe approximation was extended.
- On page 40 we emphasized that the difference between K^C and $-\log m^C$ can be inflated arbitrarily.
- On page 87, a mistake was fixed in the definition of $B^{\deg}(D)$.

SIKS DISSERTATION SERIES

INDEX

www.ingramcontent.com/pod-product-compliance
Lightning Source LLC
Chambersburg PA
CBHW070317190526
45169CB00005B/1654